BIG IDEAS MATH®
Modeling Real Life

Grade 1
Common Core Edition
Volume 1

Ron Larson
Laurie Boswell

Erie, Pennsylvania
BigIdeasLearning.com

Big Ideas Learning, LLC
1762 Norcross Road
Erie, PA 16510-3838
USA

For product information and customer support, contact Big Ideas Learning
at 1-877-552-7766 or visit us at BigIdeasLearning.com.

Cover Image
Valdis Torms, Brazhnykov Andriy/Shutterstock.com

Copyright © 2019 by Big Ideas Learning, LLC. All rights reserved.

No part of this work may be reproduced or transmitted in any form or by any means, electronic or mechanical, including, but not limited to, photocopying and recording, or by any information storage or retrieval system, without prior written permission of Big Ideas Learning, LLC, unless such copying is expressly permitted by copyright law. Address inquiries to Permissions, Big Ideas Learning, LLC, 1762 Norcross Road, Erie, PA 16510.

Big Ideas Learning and Big Ideas Math are registered trademarks of Larson Texts, Inc.

Common Core State Standards: © Copyright 2010. National Governors Association Center for Best Practices and Council of Chief State School Officers. All rights reserved.

Printed in the U.S.A.

ISBN 13: 978-1-64208-360-6

6 7 8 9 10—23 22

About the Authors

Ron Larson

Ron Larson, Ph.D., is well known as the lead author of a comprehensive program for mathematics that spans school mathematics and college courses. He holds the distinction of Professor Emeritus from Penn State Erie, The Behrend College, where he taught for nearly 40 years. He received his Ph.D. in mathematics from the University of Colorado. Dr. Larson's numerous professional activities keep him actively involved in the mathematics education community and allow him to fully understand the needs of students, teachers, supervisors, and administrators.

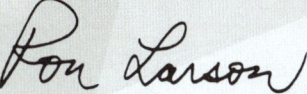

Laurie Boswell

Laurie Boswell, Ed.D., is the former Head of School at Riverside School in Lyndonville, Vermont. In addition to textbook authoring, she provides mathematics consulting and embedded coaching sessions. Dr. Boswell received her Ed.D. from the University of Vermont in 2010. She is a recipient of the Presidential Award for Excellence in Mathematics Teaching and is a Tandy Technology Scholar. Laurie has taught math to students at all levels, elementary through college. In addition, Laurie has served on the NCTM Board of Directors and as a Regional Director for NCSM. Along with Ron, Laurie has co-authored numerous math programs and has become a popular national speaker.

Dr. Ron Larson and Dr. Laurie Boswell began writing together in 1992. Since that time, they have authored over four dozen textbooks. This successful collaboration allows for one voice from Kindergarten through Algebra 2.

Contributors, Reviewers, and Research

Big Ideas Learning would like to express our gratitude to the mathematics education and instruction experts who served as our advisory panel, contributing specialists, and reviewers during the writing of *Big Ideas Math: Modeling Real Life*. Their input was an invaluable asset during the development of this program.

Contributing Specialists and Reviewers

- **Sophie Murphy**, Ph.D. Candidate, Melbourne School of Education, Melbourne, Australia
  Learning Targets and Success Criteria Specialist and Visible Learning Reviewer

- **Linda Hall**, Mathematics Educational Consultant, Edmond, OK
  Advisory Panel

- **Michael McDowell**, Ed.D., Superintendent, Ross, CA
  Project-Based Learning Specialist

- **Kelly Byrne**, Math Supervisor and Coordinator of Data Analysis, Downingtown, PA
  Advisory Panel

- **Jean Carwin**, Math Specialist/TOSA, Snohomish, WA
  Advisory Panel

- **Nancy Siddens**, Independent Language Teaching Consultant, Las Cruces, NM
  English Language Learner Specialist

- **Kristen Karbon**, Curriculum and Assessment Coordinator, Troy, MI
  Advisory Panel

- **Kery Obradovich**, K–8 Math/Science Coordinator, Northbrook, IL
  Advisory Panel

- **Jennifer Rollins**, Math Curriculum Content Specialist, Golden, CO
  Advisory Panel

- **Becky Walker**, Ph.D., School Improvement Services Director, Green Bay, WI
  Advisory Panel and Content Reviewer

- **Deborah Donovan**, Mathematics Consultant, Lexington, SC
  Content Reviewer

- **Tom Muchlinski**, Ph.D., Mathematics Consultant, Plymouth, MN
  Content Reviewer and Teaching Edition Contributor

- **Mary Goetz**, Elementary School Teacher, Troy, MI
  Content Reviewer

- **Nanci N. Smith**, Ph.D., International Curriculum and Instruction Consultant, Peoria, AZ
  Teaching Edition Contributor

- **Robyn Seifert-Decker**, Mathematics Consultant, Grand Haven, MI
  Teaching Edition Contributor

- **Bonnie Spence**, Mathematics Education Specialist, Missoula, MT
  Teaching Edition Contributor

- **Suzy Gagnon**, Adjunct Instructor, University of New Hampshire, Portsmouth, NH
  Teaching Edition Contributor

- **Art Johnson**, Ed.D., Professor of Mathematics Education, Warwick, RI
  Teaching Edition Contributor

- **Anthony Smith**, Ph.D., Associate Professor, Associate Dean, University of Washington Bothell, Seattle, WA
  Reading and Writing Reviewer

- **Brianna Raygor**, Music Teacher, Fridley, MN
  Music Reviewer

- **Nicole Dimich Vagle**, Educator, Author, and Consultant, Hopkins, MN
  Assessment Reviewer

- **Janet Graham**, District Math Specialist, Manassas, VA
  Response to Intervention and Differentiated Instruction Reviewer

- **Sharon Huber**, Director of Elementary Mathematics, Chesapeake, VA
  Universal Design for Learning Reviewer

Student Reviewers

- T.J. Morin
- Alayna Morin
- Ethan Bauer
- Emery Bauer
- Emma Gaeta
- Ryan Gaeta
- Benjamin SanFrotello
- Bailey SanFrotello
- Samantha Grygier
- Robert Grygier IV
- Jacob Grygier
- Jessica Urso
- Ike Patton
- Jake Lobaugh
- Adam Fried
- Caroline Naser
- Charlotte Naser

Research

Ron Larson and Laurie Boswell used the latest in educational research, along with the body of knowledge collected from expert mathematics instructors, to develop the *Modeling Real Life* series. The pedagogical approach used in this program follows the best practices outlined in the most prominent and widely accepted educational research, including:

- *Visible Learning*
  John Hattie © 2009
- *Visible Learning for Teachers*
  John Hattie © 2012
- *Visible Learning for Mathematics*
  John Hattie © 2017
- *Principles to Actions: Ensuring Mathematical Success for All*
  NCTM © 2014
- *Adding It Up: Helping Children Learn Mathematics*
  National Research Council © 2001
- *Mathematical Mindsets: Unleashing Students' Potential through Creative Math, Inspiring Messages and Innovative Teaching*
  Jo Boaler © 2015
- *What Works in Schools: Translating Research into Action*
  Robert Marzano © 2003
- *Classroom Instruction That Works: Research-Based Strategies for Increasing Student Achievement*
  Marzano, Pickering, and Pollock © 2001
- *Principles and Standards for School Mathematics*
  NCTM © 2000
- *Rigorous PBL by Design: Three Shifts for Developing Confident and Competent Learners*
  Michael McDowell © 2017
- Common Core State Standards for Mathematics
  National Governors Association Center for Best Practices and Council of Chief State School Officers © 2010
- *Universal Design for Learning Guidelines*
  CAST © 2011
- Rigor/Relevance Framework®
  International Center for Leadership in Education
- *Understanding by Design*
  Grant Wiggins and Jay McTighe © 2005
- Achieve, ACT, and The College Board
- *Elementary and Middle School Mathematics: Teaching Developmentally*
  John A. Van de Walle and Karen S. Karp © 2015
- *Evaluating the Quality of Learning: The SOLO Taxonomy*
  John B. Biggs & Kevin F. Collis © 1982
- *Unlocking Formative Assessment: Practical Strategies for Enhancing Students' Learning in the Primary and Intermediate Classroom*
  Shirley Clarke, Helen Timperley, and John Hattie © 2004
- *Formative Assessment in the Secondary Classroom*
  Shirley Clarke © 2005
- *Improving Student Achievement: A Practical Guide to Assessment for Learning*
  Toni Glasson © 2009

Standards for Mathematical Practice

1. **Make sense of problems and persevere in solving them.**
   - Multiple representations are presented to help students move from concrete to representative and into abstract thinking.
   - In *Modeling Real Life* examples and exercises, students MAKE SENSE OF PROBLEMS using problem-solving strategies, such as drawing a picture, circling knowns, and underlining unknowns.

2. **Reason abstractly and quantitatively.**
   - Visual problem-solving models help students create a coherent representation of the problem.
   - *Explore and Grows* allow students to investigate concepts to understand the REASONING behind the rules.
   - Exercises encourage students to apply NUMBER SENSE and explain and justify their REASONING.

3. **Construct viable arguments and critique the reasoning of others.**
   - *Explore and Grows* help students make conjectures, use LOGIC, and CONSTRUCT ARGUMENTS to support their conjectures.
   - Exercises, such as *You Be The Teacher* and *Which One Doesn't Belong?*, provide students the opportunity to CRITIQUE REASONING.

4. **Model with mathematics.**
   - Real-life situations are translated into pictures, diagrams, tables, equations, or graphs to help students analyze relations and to draw conclusions.
   - Real-life problems are provided to help students apply the mathematics they are learning to everyday life.
   - MODELING REAL LIFE examples and exercises help students see that math is used across content areas, other disciplines, and in their own experiences.

5. **Use appropriate tools strategically.**
   - Students can use a variety of hands-on manipulatives to solve problems throughout the program.
   - A variety of tools, such as number lines and pattern blocks, manipulatives, and digital tools, are available as students CHOOSE TOOLS and consider how to approach a problem.

6. **Attend to precision.**
   - PRECISION exercises encourage students to formulate consistent and appropriate reasoning.
   - Cooperative learning opportunities support precise communication.

7. **Look for and make use of structure.**
   - *Learning Targets* and *Success Criteria* at the start of each chapter and lesson help students understand what they are going to learn.
   - *Explore and Grows* provide students the opportunity to see PATTERNS and STRUCTURE in mathematics.
   - Real-life problems help students use the STRUCTURE of mathematics to break down and solve more difficult problems.

8. **Look for and express regularity in repeated reasoning.**
   - Opportunities are provided to help students make generalizations through REPEATED REASONING.
   - Students are continually encouraged to check for reasonableness in their solutions.

The colored words above are used throughout the program to indicate exercises that correlate to the Standards for Mathematical Practice.

Achieve the Core

Meeting Proficiency

As standards shift to prepare students for college and careers, the importance of focus, coherence, and rigor continues to grow.

FOCUS — *Big Ideas Math: Modeling Real Life* emphasizes a narrower and deeper curriculum, ensuring students spend their time on the major topics of each grade.

COHERENCE — The program was developed around coherent progressions from Kindergarten through eighth grade, guaranteeing students develop and progress their foundational skills through the grades while maintaining a strong focus on the major topics.

RIGOR — *Big Ideas Math: Modeling Real Life* uses a balance of procedural fluency, conceptual understanding, and real-life applications. Students develop conceptual understanding in every *Explore and Grow*, continue that development through the lesson while gaining procedural fluency during the *Think and Grow*, and then tie it all together with *Think and Grow: Modeling Real Life*. Every set of practice problems reflects this balance, giving students the rigorous practice they need to be college- and career-ready.

Major Topics in Grade 1

Operations and Algebraic Thinking
- Represent and solve problems involving addition and subtraction.
- Understand and apply properties of operations and the relationship between addition and subtraction.
- Add and subtract within 20.
- Work with addition and subtraction equations.

Number and Operations in Base Ten
- Extending the counting sequence.
- Understand place value.
- Use place value understanding and properties of operations to add and subtract.

Measurement and Data
- Measure lengths indirectly and by iterating length units.

Use the color-coded Table of Contents to determine where the major topics, supporting topics, and additional topics occur throughout the curriculum.

- 🟩 Major Topic
- 🟦 Supporting Topic
- 🟨 Additional Topic

1 Addition and Subtraction Situations

	Vocabulary	2
■	1.1 Addition: *Add To*	3
■	1.2 Solve *Add To* Problems	9
■	1.3 Solve *Put Together* Problems	15
■	1.4 Solve *Put Together* Problems with Both Addends Unknown	21
■	1.5 Solve *Take From* Problems	27
■	1.6 Solve *Compare* Problems: More	33
■	1.7 Solve *Compare* Problems: Fewer	39
■	1.8 Solve *Add To* Problems with Change Unknown	45
■	1.9 Connect *Put Together* and *Take Apart* Problems	51
	Performance Task: Birds	57
	Game: Three in a Row	58
	Chapter Practice	59

2 Fluency and Strategies within 10

	Vocabulary	64
■	2.1 Add 0	65
■	2.2 Subtract 0 and Subtract All	71
■	2.3 Add and Subtract 1	77
■	2.4 Add Doubles from 1 to 5	83
■	2.5 Use Doubles	89
■	2.6 Add in Any Order	95
■	2.7 Count On to Add	101
■	2.8 Count Back to Subtract	107
■	2.9 Use Addition to Subtract	113
	Performance Task: Flowers	119
	Game: Add or Subtract	120
	Chapter Practice	121

■ Major Topic
■ Supporting Topic
■ Additional Topic

3 More Addition and Subtraction Situations

	Vocabulary	126
■ 3.1	Solve *Add To* Problems with Start Unknown	127
■ 3.2	Solve *Take From* Problems with Change Unknown	133
■ 3.3	Solve *Take From* Problems with Start Unknown	139
■ 3.4	*Compare* Problems: Bigger Unknown	145
■ 3.5	*Compare* Problems: Smaller Unknown	151
■ 3.6	True or False Equations	157
■ 3.7	Find Numbers That Make 10	163
■ 3.8	Fact Families	169
	Performance Task: Baking	175
	Game: Number Land	176
	Chapter Practice	177
	Cumulative Practice	181

Number Land

To Play: Put the Addition and Subtraction Cards in a pile. Start at Newton. Take turns drawing a card and moving your piece to the missing number in the equation. Repeat this process until a player gets back to Newton.

Add Numbers within 20

	Vocabulary	186
4.1	Add Doubles from 6 to 10	187
4.2	Use Doubles within 20	193
4.3	Count On to Add within 20	199
4.4	Add Three Numbers	205
4.5	Add Three Numbers by Making a 10	211
4.6	Add 9	217
4.7	Make a 10 to Add	223
4.8	Problem Solving: Addition within 20	229
	Performance Task: Weather	235
	Game: Roll and Cover	236
	Chapter Practice	237

Subtract Numbers within 20

	Vocabulary	242
5.1	Count Back to Subtract within 20	243
5.2	Use Addition to Subtract within 20	249
5.3	Subtract 9	255
5.4	Get to 10 to Subtract	261
5.5	More True or False Equations	267
5.6	Make True Equations	273
5.7	Problem Solving: Subtraction within 20	279
	Performance Task: Bees	285
	Game: Three in a Row: Subtraction	286
	Chapter Practice	287

■ Major Topic
■ Supporting Topic
■ Additional Topic

x

6 Count and Write Numbers to 120

	Vocabulary	292
6.1	Count to 120 by Ones	293
6.2	Count to 120 by Tens	299
6.3	Compose Numbers 11 to 19	305
6.4	Tens	311
6.5	Tens and Ones	317
6.6	Make Quick Sketches	323
6.7	Understand Place Value	329
6.8	Write Numbers in Different Ways	335
6.9	Count and Write Numbers to 120	341
	Performance Task: Fundraiser	347
	Game: Drop and Build	348
	Chapter Practice	349

7 Compare Two-Digit Numbers

	Vocabulary	354
7.1	Compare Numbers 11 to 19	355
7.2	Compare Numbers	361
7.3	Compare Numbers Using Place Value	367
7.4	Compare Numbers Using Symbols	373
7.5	Compare Numbers Using a Number Line	379
7.6	1 More, 1 Less; 10 More, 10 Less	385
	Performance Task: Toy Drive	391
	Game: Number Boss	392
	Chapter Practice	393
	Cumulative Practice	397

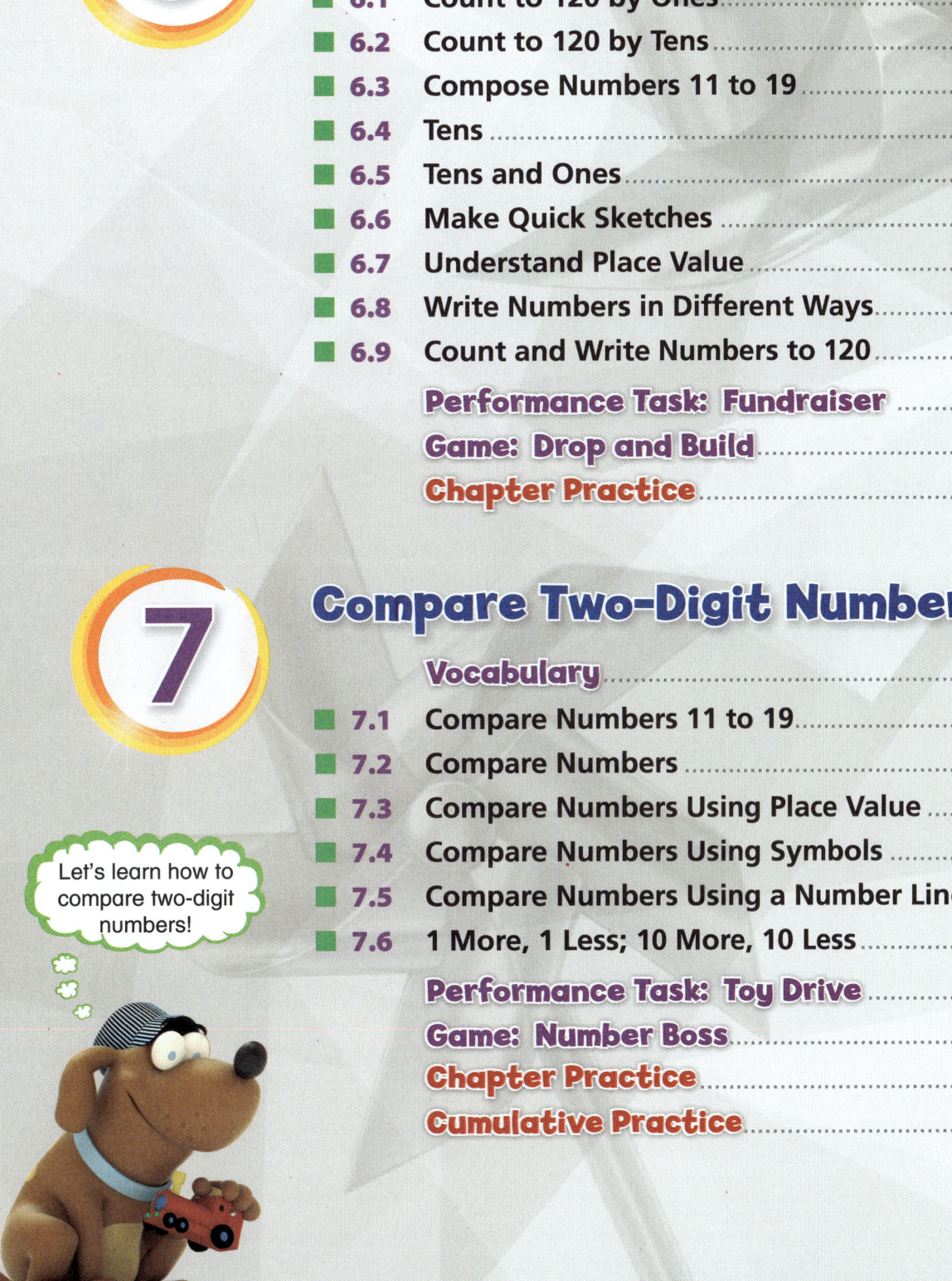

Let's learn how to compare two-digit numbers!

Add and Subtract Tens

	Vocabulary	402
8.1	Mental Math: 10 More	403
8.2	Mental Math: 10 Less	409
8.3	Add Tens	415
8.4	Add Tens Using a Number Line	421
8.5	Subtract Tens	427
8.6	Subtract Tens Using a Number Line	433
8.7	Use Addition to Subtract Tens	439
8.8	Add Tens to a Number	445
	Performance Task: Motion	451
	Game: 10 More or 10 Less	452
	Chapter Practice	453

Add Two-Digit Numbers

	Vocabulary	458
9.1	Add Tens and Ones	459
9.2	Add Tens and Ones Using a Number Line	465
9.3	Make a 10 to Add	471
9.4	Add Two-Digit Numbers	477
9.5	Practice Addition Strategies	483
9.6	Problem Solving: Addition	489
	Performance Task: Games	495
	Game: Race for 100	496
	Chapter Practice	497

■ Major Topic
■ Supporting Topic
■ Additional Topic

10 Measure and Compare Lengths

	Vocabulary	502
■	10.1 Order Objects by Length	503
■	10.2 Compare Lengths Indirectly	509
■	10.3 Measure Lengths	515
■	10.4 Measure More Lengths	521
■	10.5 Solve *Compare* Problems Involving Length	527
	Performance Task: Maps	533
	Game: Fish Measurement	534
	Chapter Practice	535

Think and Grow

Use color tiles to **measure** lengths of objects.

Do not leave gaps or overlap the tiles.

length unit

about __4__ color tiles

xiii

11 Represent and Interpret Data

	Vocabulary	540
■	11.1 Sort and Organize Data	541
■	11.2 Read and Interpret Picture Graphs	547
■	11.3 Read and Interpret Bar Graphs	553
■	11.4 Represent Data	559
■	11.5 Solve Problems Involving Data	565
	Performance Task: Eye Color	571
	Game: Spin and Graph	572
	Chapter Practice	573
	Cumulative Practice	577

12 Tell Time

	Vocabulary	582
■	12.1 Tell Time to the Hour	583
■	12.2 Tell Time to the Half Hour	589
■	12.3 Tell Time to the Hour and Half Hour	595
■	12.4 Tell Time Using Analog and Digital Clocks	601
	Performance Task: Field Trip	607
	Game: Time Flip and Find	608
	Chapter Practice	609

■ Major Topic
■ Supporting Topic
■ Additional Topic

13 Two- and Three-Dimensional Shapes

Vocabulary	612
13.1 Sort Two-Dimensional Shapes	613
13.2 Describe Two-Dimensional Shapes	619
13.3 Combine Two-Dimensional Shapes	625
13.4 Create More Shapes	631
13.5 Take Apart Two-Dimensional Shapes	637
13.6 Sort Three-Dimensional Shapes	643
13.7 Describe Three-Dimensional Shapes	649
13.8 Combine Three-Dimensional Shapes	655
13.9 Take Apart Three-Dimensional Shapes	661
Performance Task: Sandcastles	667
Game: Shape Roll and Build	668
Chapter Practice	669

14 Equal Shares

Vocabulary	674
14.1 Equal Shares	675
14.2 Partition Shapes into Halves	681
14.3 Partition Shapes into Fourths	687
Performance Task: Picnic	693
Game: Three in a Row: Equal Shares	694
Chapter Practice	695
Cumulative Practice	697
Glossary	A1
Index	A11
Reference Sheet	A25

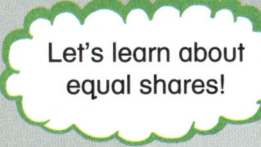

Let's learn about equal shares!

1 Addition and Subtraction Situations

- How many birds do you see? How many are flying away?
- How many birds are left on the wire?

Chapter Learning Target:
Understand addition.

Chapter Success Criteria:
- I can identify a group of objects.
- I can describe numbers as a group.
- I can write an addition equation and a subtraction equation.
- I can model addition and subtraction.

1 Vocabulary

Organize It

Review Words
equal sign
minus sign
plus sign

Use the review words to complete the graphic organizer.

```
[ plus sign ]              [ minus sign ]
     ↓                          ↓
   2 + 3 = 5                 5 − 3 = 2
        ↑                      ↑
          [ equal sign ]
```

Define It

Use your vocabulary cards to match.

1. more

2. fewer

3. addition equation

4. subtraction equation

9 − 5 = 4

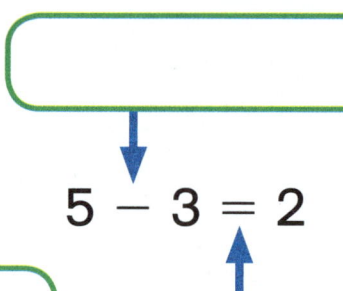

4 + 5 = 9

Chapter 1 Vocabulary Cards

add	addend
addition equation	**difference**
equals	**fewer**
minus	**more**

4 + 3 = 7	● ● + ● ● ● = ● ● ● ● ● ● ● ● ● ● ● ● ● ● 2 + 4 = 6
8 − 3 = 5	4 + 5 = 9
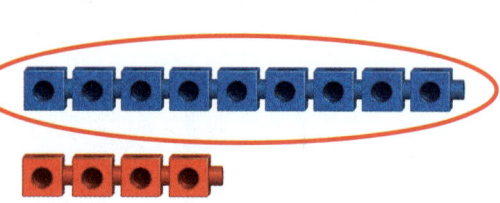	8 + 2 = 10 8 plus 2 equals 10
(blue cubes circled) / (red cubes)	3 − 1 3 minus 1

Chapter 1 Vocabulary Cards

| part | part-part-whole model |

| plus | subtract |

| subtraction equation | sum |

| whole | |

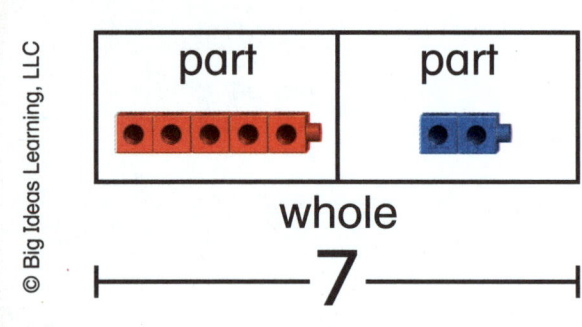

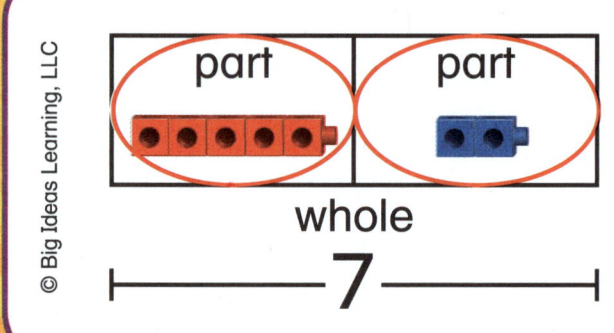

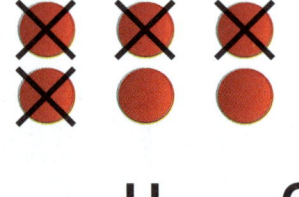

6 − 4 = 2

2 + 1
2 plus 1

5 + 3 = 8

9 − 5 = 4

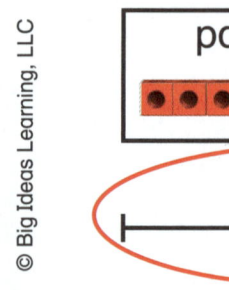
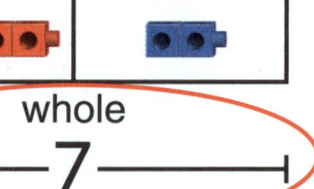

Name _____

Addition: Add To

1.1

Learning Target: Add to a group of objects and write an addition equation.

Explore and Grow

Use linking cubes to model the story.

There are 4 🐻. 2 more 🐻 join them.

How many 🐻 are there now?

_____

Chapter 1 | Lesson 1

three 3

 **Think and Grow**

There are 3 🐶.

2 more 🐶 join them.

Now there are __5__ 🐶.

An addition sentence is also called an *addition equation*.

addition equation

__3__ + __2__ = __5__

Show and Grow I can do it!

1.

There are 5 🐱.

1 more 🐱 joins them.

Now there are ____ 🐱.

Addition equation: ____ + ____ = ____

Name _____

✓ Apply and Grow: Practice

2.

There are 3 🦀.

6 more 🦀 join them.

Now there are ___ 🦀. ___ + ___ = ___

3.

There are 8 🐷.

2 more 🐷 join them.

Now there are ___ 🐷. ___ + ___ = ___

4. **DIG DEEPER!** Complete the picture and the story.

There are ___ 🦉.

3 more 🦉 join them.

Now there are ___ 🦉. ___ + ___ = ___

Chapter 1 | Lesson 1

Think and Grow: Modeling Real Life

There are 5 🐟. 3 more 🐟 join them. How many 🐟 are there now?

Draw a picture:

Addition equation:

_____

Show and Grow — I can think deeper!

5. There are 2 🐢. 7 more 🐢 join them. How many 🐢 are there now?

Draw a picture:

Addition equation:

_____

Name _____

Practice

Learning Target: Add to a group of objects and write an addition equation.

There are 2 🐑.

5 more 🐑 join them.

Now there are __7__ 🐑.

Addition equation:

__2__ + __5__ = __7__

1.

   There are 6 🦊.

   1 more 🦊 joins them.

   Now there are ____ 🦊.

   ___ + ___ = ___

2.

   There are 7 🐞.

   3 more 🐞 join them.

   Now there are ____ 🐞.

   ___ + ___ = ___

Chapter 1 | Lesson 1   seven 7

3. **DIG DEEPER!** Complete the picture and the story.

There are ____ 🦜.

2 more 🦜 join them.

Now there are ____ 🦜.

____ + ____ = ____

4. **Modeling Real Life** There are 2 🦆. 7 more 🦆 join them. How many 🦆 are there now?

____

Review & Refresh

Write the number of objects.

5.

____

6.

____

8   eight

Name _____

Solve *Add To* Problems 1.2

Learning Target: Solve *add to* word problems.

Use linking cubes to model the story.

There are 5 🎩.

You add 2 more 🎩.

How many 🎩 are there now?

____ 🎩

Chapter 1 | **Lesson 2**

nine 9

Think and Grow

You have 3 🟥.

You **add** 4 more 🟥.

How many 🟥 do you have now?

$$\underline{3} \quad + \quad \underline{4} \quad = \quad \underline{7}$$
addend   **plus**   **addend**   **equals**   **sum**

$$\underline{7} \text{ 🟥}$$

Show and Grow — I can do it!

1. You have 4 🟣.

   You find 2 more 🟣.

   How many 🟣 do you have now?

   Addition equation:

   ___ + ___ = ___

   ___ 🟣

2. There are 4 🐰.

   4 more 🐰 join them.

   How many 🐰 are there now?

   Addition equation:

   ___ + ___ = ___

   ___ 🐰

Name _____

✓ Apply and Grow: Practice

3. You have 2 📗.

You buy 2 more 📗.

How many 📗 do you have now?

____ + ____ = ____

____ 📗

---

4. There are 3 🐜.

5 more 🐜 join them.

How many 🐜 are there now?

____ + ____ = ____

____ 🐜

---

5. You eat 9 🟢.

You eat 1 more 🟢.

How many 🟢 do you eat?

____ + ____ = ____

____ 🟢

---

6. **DIG DEEPER!** Complete the picture and the story.

There are 4 🐌.

3 more 🐌 join them.

Now there are ____ 🐌.

____ + ____ = ____

Chapter 1 | Lesson 2

eleven 11

Think and Grow: Modeling Real Life

You have 3 🍎. You buy 7 🍎.

How many 🍎 do you have now?

Draw a picture:

Addition equation:

_____ 🍎

Show and Grow — I can think deeper!

7. You have 6 🟦. Your friend gives you 3 🟦.
How many 🟦 do you have now?

Draw a picture:

Addition equation:

_____ 🟦

Name _____

Practice 1.2

Learning Target: Solve *add to* word problems.

You have 1 🟥.

You add 4 more 🟥.

How many 🟥 do you have now?

<u> 1 </u> + <u> 4 </u> = <u> 5 </u>

<u> 5 </u> 🟥

1. You have 2 ✏️.

   You buy 3 more ✏️.

   How many ✏️ do you have now?

   ___ + ___ = ___

   ___ ✏️

2. There are 6 🌷.

   You add 4 more 🌷.

   How many 🌷 are there now?

   ___ + ___ = ___

   ___ 🌷

3. There are 5 🦋.

   4 more 🦋 join them.

   How many 🦋 are there now?

   ___ + ___ = ___

   ___ 🦋

Chapter 1 | Lesson 2

thirteen 13

4. **DIG DEEPER!** Complete the picture and the story.

There are 7 🐸.

3 more 🐸 join them.

Now there are ___ 🐸.

___ + ___ = ___

5. **Modeling Real Life** You have 2 ⚽. You buy 4 ⚽. How many ⚽ do you have now?

___

Review & Refresh

6. Use the picture to complete the number bond.

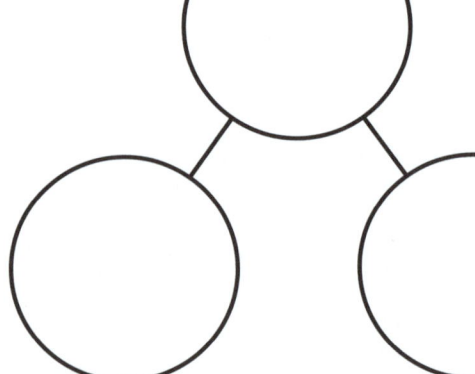

14   fourteen

Name _____

Learning Target: Solve *put together* word problems.

Solve *Put Together* Problems 1.3

Explore and Grow

Use counters to model the story.

You have 7 🤖 and 2 🚗. How many toys do you have in all?

_____ toys

Chapter 1 | Lesson 3

fifteen 15

Think and Grow

You have 5 🖍️ and 2 🖍️. How many crayons do you have in all?

I can use a part-part-whole model to show the parts and the whole.

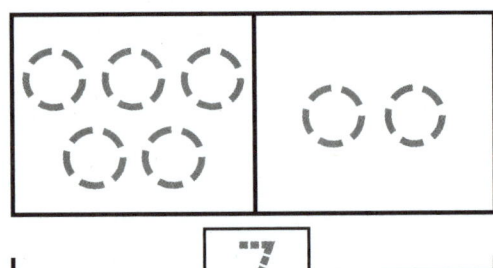

$$\underline{5} + \underline{2} = \underline{7}$$
part   part   whole

part-part-whole model

$\underline{7}$ crayons

Show and Grow — I can do it!

1. You have 3 🏀 and 1 ⚽. How many balls do you have in all?

   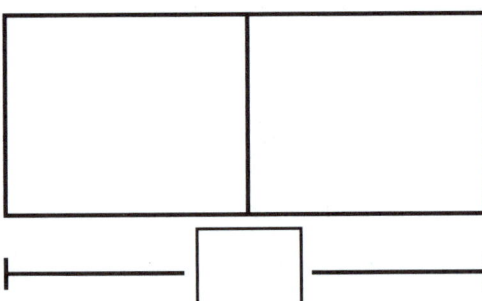

   $\underline{} + \underline{} = \underline{}$

   $\underline{}$ balls

2. There are 2 🐟 and 6 🐟. How many fish are there in all?

   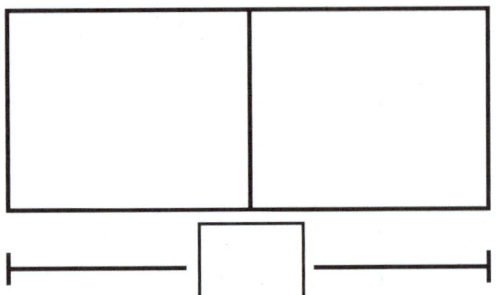

   $\underline{} + \underline{} = \underline{}$

   $\underline{}$ fish

Name _____

✓ Apply and Grow: Practice

3. You have 1 🐕 and 5 🐈. How many pets do you have in all?

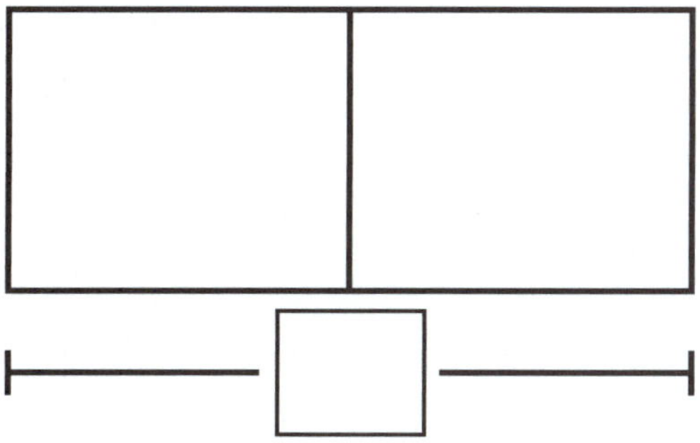

___ + ___ = ___

___ pets

4. There are 4 🐎 and 6 🐔. How many animals are there in all?

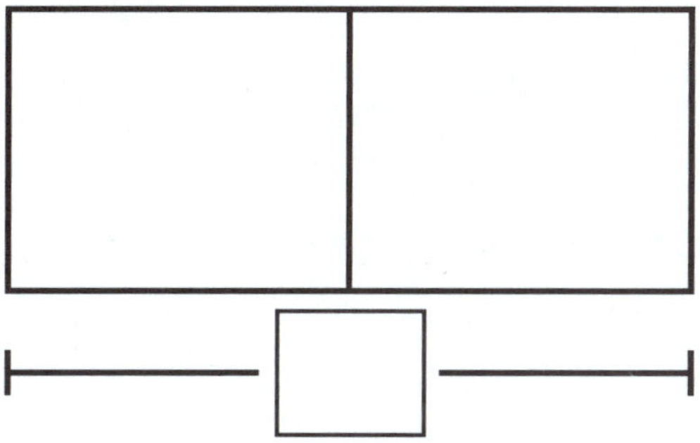

___ + ___ = ___

___ animals

5. **Number Sense** Circle two groups of bugs to match the addition problem. Then find the sum.

3 + 5 = ___

Chapter 1 | Lesson 3   seventeen   17

Think and Grow: Modeling Real Life

You have 3 🌷 and 4 🌷. Your friend has 8 flowers. Who has more flowers?

Model:

Addition equation: _____

Who has more flowers?   You   Friend

Show and Grow I can think deeper!

6. You have 4 🍂 and 5 🍃. Your friend has 7 leaves. Who has more leaves?

Model:

Addition equation: _____

Who has more leaves?   You   Friend

18 eighteen

Name _____

Practice 1.3

Learning Target: Solve *put together* word problems.

You have 4 🟥 and 1 🟦. How many linking cubes do you have in all?

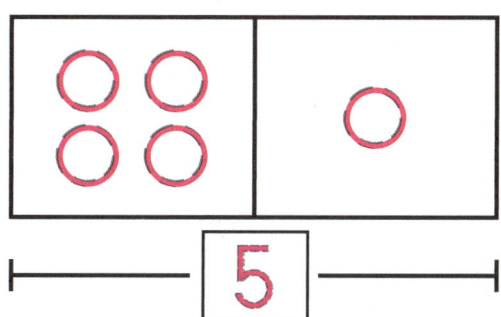

__4__ + __1__ = __5__

__5__ cubes

1. You buy 1 🍍 and 7 🍌. How many pieces of fruit do you buy in all?

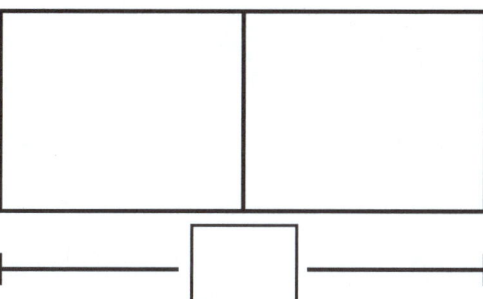

___ + ___ = ___

___ pieces of fruit

2. There are 5 🐦 and 5 🐦. How many birds are there in all?

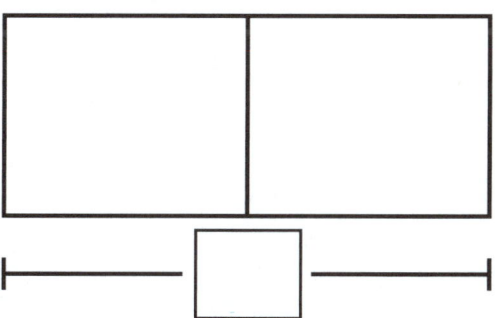

___ + ___ = ___

___ birds

Chapter 1 | Lesson 3

nineteen 19

3. **Number Sense** Circle two groups of hats to match the addition problem. Then find the sum.

6 + 1 = ___

4. **Modeling Real Life** You have 4  and 5 . Your friend has 8 cars. Who has more cars?

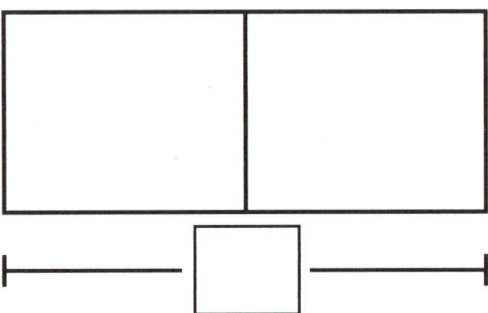

Who has more cars?   You   Friend

Review & Refresh

5. Write the number for each type of pepper. Write the number for the whole.

___ ___ ___

Name _____

Learning Target: Find addends for a given sum.

Solve *Put Together* Problems with Both Addends Unknown 1.4

Explore and Grow

Use linking cubes to model the story.

There are 10 . Some are on the court and some are on the rack.

____ on the court

____ on the rack

Chapter 1 | Lesson 4

Think and Grow

There are 6 🦋. Some are inside a jar. Some more are outside the jar. Draw the 🦋.

Addition equation: __6__ = __2__ + __4__

Show and Grow — I can do it!

1. There are 7 🍒. Some are on a tree. Some more are on the ground. Draw the 🍒.

Addition equation: ____ = ____ + ____

Name _____

✓ Apply and Grow: Practice

2. There are 5 . Some are on a couch. Some more are on a rug. Draw the .

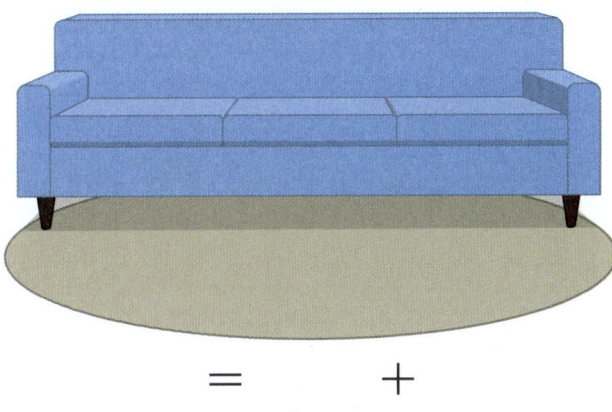

___ = ___ + ___

3. There are 8 🐍. Some are in a pond. Some more are in the grass. Draw two different pictures to show the 🐍.

___ = ___ + ___   ___ = ___ + ___

4. **DIG DEEPER!** Newton has 4 toy trucks. Some are red. The rest are blue. Write two different addition equations to describe his trucks.

___ = ___ + ___   ___ = ___ + ___

Chapter 1 | Lesson 4   twenty-three **23**

Think and Grow: Modeling Real Life

You have 7 ribbons. Some are blue. The rest are red. You have more blue ribbons than red ribbons. How many blue and red ribbons can you have?

    2 blue   5 red   4 blue   3 red

    6 blue   1 red   3 blue   4 red

Show how you know:

Show and Grow — I can think deeper!

5. There are 10 kites. Some are green. The rest are yellow. There are more yellow kites than green kites. How many green and yellow kites can there be?

    4 green   6 yellow   6 green   4 yellow

    5 green   5 yellow   3 green   7 yellow

Show how you know:

Name _____

Practice 1.4

Learning Target: Find addends for a given sum.

There are 4 . Some are inside a box. Some more are outside the box. Draw the .

Addition equation: __4__ = __1__ + __3__

1. There are 6 🐜. Some are on a hill. Some more are in the grass. Draw the 🐜.

___ = ___ + ___

2. There are 8 🐸. Some are in the water. Some more are on the dirt. Draw the 🐸.

___ = ___ + ___

3. There are 5 ⚽. Some are inside a bin. Some more are outside the bin. Draw two different pictures to show the ⚽.

 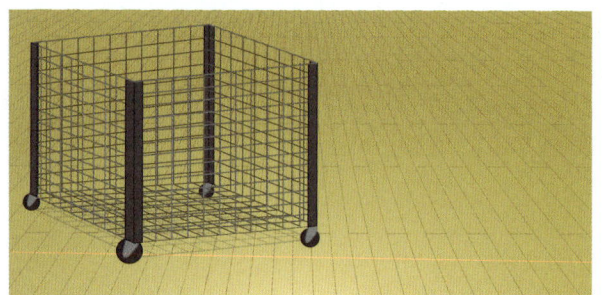

___ = ___ + ___   ___ = ___ + ___

4. **Modeling Real Life** You have 9 hats. Some are blue. The rest are red. You have more red hats than blue hats. How many red and blue hats can you have?

    4 red   5 blue   6 red   3 blue

    3 red   6 blue   7 red   2 blue

Show how you know:

Review & Refresh

Use the picture to write a subtraction sentence.

5.

   ___ − ___ = ___

6.

   ___ − ___ = ___

Name _____

Learning Target: Solve *take from* word problems.

Solve *Take From* Problems

Explore and Grow

Use linking cubes to model the story.

There are 6 🐰. 3 🐰 hop away.

How many 🐰 are left?

_____ 🐰

Chapter 1 | Lesson 5

twenty-seven 27

Think and Grow

You have 4 cube.

You take away 1 cube.

How many cube do you have left?

subtraction equation

$$4 - 1 = 3$$

minus **difference**

___3___ cube

Subtract to find the difference.

Show and Grow — I can do it!

1. There are 6 .

   2  jump away.

   How many are left?

   Subtraction equation:

   ___ − ___ = ___

   ___ monkey

2. There are 8 .

   A monkey eats 5 banana.

   How many banana are left?

   Subtraction equation:

   ___ − ___ = ___

   ___

Name _____

✓ Apply and Grow: Practice

3. There are 7

3  walk away.

How many 🐜 are left?

___ − ___ = ___

___

---

4. There are 8 🐦.

6 🐦 fly away.

How many 🐦 are left?

___ − ___ = ___

___

---

5. 10 🐱 play.

5 🐱 run away.

How many 🐱 are left?

___ − ___ = ___

___

---

6. **MP Precision** Find the difference. Draw a picture to match.

$$7 - 4 = \underline{}$$

Think and Grow: Modeling Real Life

7 students play tag. 2 of them leave. How many students are left?

Which equation matches the story?

$9 - 7 = 2$ $7 - 2 = 5$ $7 - 5 = 2$

Show how you know:

_____ students are left.

Show and Grow I can think deeper!

7. You have 9 coins. You give 3 coins away. How many coins are left?

Which equation matches the story?

$9 - 3 = 6$ $6 - 3 = 3$ $9 - 6 = 3$

Show how you know:

_____ coins are left.

Name _____

Practice 1.5

Learning Target: Solve *take from* word problems.

You have 6 🟦.

You take away 4 🟦.

How many 🟦 do you have left?

Subtraction equation:

$\underline{6} - \underline{4} = \underline{2}$

$\underline{2}$ 🟦

1. There are 5 🐅.

   3 🐅 leave.

   How many 🐅 are left?

   ___ − ___ = ___

   ___ 🐅

2. There are 7 🐟.

   A cat eats 5 🐟.

   How many 🐟 are left?

   ___ − ___ = ___

   ___ 🐟

3. You have 10 🎾.

   You give away 10 🎾.

   How many 🎾 do you have left?

   ___ − ___ = ___

   ___ 🎾

Chapter 1 | Lesson 5

4. **Precision** Find the difference. Draw a picture to match.

$$8 - 2 = \underline{}$$

---

5. **Modeling Real Life** 6 students play soccer. 4 of them leave. How many students are left?

   Which equation matches the story?

   $6 - 2 = 4$ $\qquad$ $8 - 6 = 2$ $\qquad$ $6 - 4 = 2$

   Show how you know:

   $\underline{}$ students are left.

Review & Refresh

6. Write the numbers of dogs and bones. Circle the number that is greater than the other number.

32 thirty-two

Name _____

Learning Target: Solve *compare* word problems by finding how many more.

Solve *Compare* Problems: More 1.6

 Explore and Grow

Use counters to model the story.

Newton has 4 apples. Descartes has 6 apples. Who has more apples? How many more?

 has ___ more apples.

(Circle one.)

Chapter 1 | Lesson 6 — thirty-three 33

Think and Grow

Newton has 5 balloons. Descartes has 3 balloons. How many **more** balloons does Newton have?

Subtraction equation: $\underline{5} - \underline{3} = \underline{2}$

$\underline{2}$ more balloons

Show and Grow I can do it!

1. You have 7 books. Your friend has 3 books. How many more books do you have?

Subtraction equation: ___ − ___ = ___

 more books

Apply and Grow: Practice

2. Your friend has 6 crayons. You have 3 crayons. How many more crayons does your friend have?

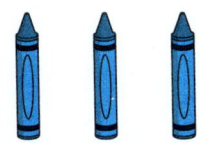

___ – ___ = ___

___ more crayons

3. There are 9 purple flowers and 5 yellow flowers. Draw the missing flowers. How many more purple flowers are there?

___ – ___ = ___

___ more purple flowers

4. You have 2 yellow buckets. Your friend has 6 blue buckets. Draw the missing buckets. How many more buckets does your friend have?

___ – ___ = ___

___ more buckets

Chapter 1 | Lesson 6

Think and Grow: Modeling Real Life

You have 2 dog puppets. You have more cat puppets than dog puppets. How many more cat puppets do you have?

Draw a picture:

Subtraction equation:

_____ more cat puppets

Show and Grow   I can think deeper!

5. You have 9 toy cars. You have more cars than trucks. How many more cars do you have?

Draw a picture:

Subtraction equation:

_____ more toy cars

36   thirty-six

Name _____

Practice 1.6

Learning Target: Solve *compare* word problems by finding how many more.

You have 8 marbles. Your friend has 5 marbles. How many more marbles do you have?

Subtraction equation: __8__ − __5__ = __3__

__3__ more marbles

1. There are 4 blue scarves and 3 green scarves. How many more blue scarves are there?

___ − ___ = ___

___ more blue scarf

2. You have 6 strawberries and 2 blueberries. Draw the missing fruit. How many more strawberries do you have?

___ − ___ = ___

___ more strawberries

Chapter 1 | Lesson 6   thirty-seven 37

3. You have 2 blue towels. Your friend has 7 red towels. Draw the missing towels. How many more towels does your friend have?

____ more towels

4. Modeling Real Life You have 3 toy bears. You have more yo-yos than toy bears. How many more yo-yos do you have?

____ more yo-yos

Review & Refresh

5. Write the number of each type of sea creature. Draw a line through the number that is less than the other number.

Name _____

Learning Target: Solve *compare* word problems by finding how many fewer.

Solve *Compare* Problems: Fewer

Explore and Grow

Use counters to model the story.

Newton has 7 games. Descartes has 5 games. Who has fewer games? How many fewer?

 has ____ fewer games.

(circle one)

Chapter 1 | Lesson 7

thirty-nine 39

 **Think and Grow**

Newton has 4 balls. Descartes has 7 balls. How many **fewer** balls does Newton have?

Subtraction equation: $7 - 4 = 3$

___3___ fewer balls

Show and Grow   I can do it!

1. There are 2 green pinwheels and 8 yellow pinwheels. How many fewer green pinwheels are there?

Subtraction equation: ____ − ____ = ____

____ fewer green pinwheels

Name _____

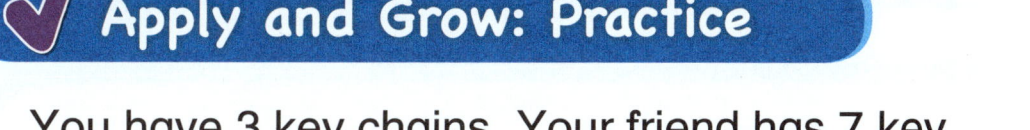

Apply and Grow: Practice

2. You have 3 key chains. Your friend has 7 key chains. How many fewer key chains do you have?

___ − ___ = ___

___ fewer key chains

3. There are 6 brown horses and 10 gray horses. Draw the missing horses. How many fewer brown horses are there?

___ − ___ = ___

___ fewer brown horses

4. You have 8 red trains. Your friend has 5 orange trains. Draw the missing trains. How many fewer trains does your friend have?

___ − ___ = ___

___ fewer trains

Chapter 1 | Lesson 7

Think and Grow: Modeling Real Life

There are 10 hens. There are fewer chicks than hens. How many fewer chicks are there?

Draw a picture:

Subtraction equation:

_____ fewer chicks

Show and Grow — I can think deeper!

5. There are 4 pigs. There are fewer pigs than piglets. How many fewer pigs are there?

Draw a picture:

Subtraction equation:

_____ fewer pigs

42 forty-two

Name _____

Practice 1.7

Learning Target: Solve *compare* word problems by finding how many fewer.

Descartes has 2 blocks. Newton has 5 blocks. How many fewer blocks does Descartes have?

Subtraction equation: __5__ − __2__ = __3__

__3__ fewer blocks

1. There are 7 black bears and 8 brown bears. How many fewer black bears are there?

___ − ___ = ___

___ fewer black bear

2. You have 4 purple blocks. Your friend has 9 yellow blocks. Draw the missing blocks. How many fewer blocks do you have?

___ − ___ = ___

___ fewer blocks

Chapter 1 | Lesson 7

forty-three 43

3. There are 7 orange kittens and 2 black kittens. Draw the missing kittens. How many fewer black kittens are there?

____ − ____ = ____

____ fewer black kittens

4. **Modeling Real Life** There are 9 lions. There are fewer cubs than lions. How many fewer cubs are there?

____ fewer cubs

Review & Refresh

Draw more counters to show how many in all. Use the ten frame to complete the addition equation.

5.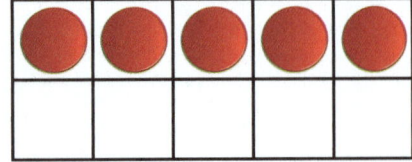

5 + ____ = 7

6.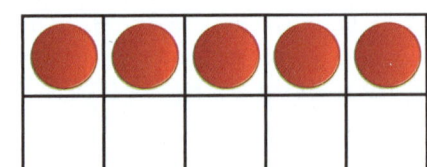

5 + ____ = 9

Name _____

Learning Target: Solve *add to* word problems that involve a missing addend.

Solve *Add To* Problems with Change Unknown

Explore and Grow

Use counters to model the story.

5 people are on a subway. Some more people get on at the subway stop. Now there are 8 people on the subway. How many people got on at the subway stop?

_____ people

Chapter 1 | Lesson 8

forty-five 45

 **Think and Grow**

You have 6 pennies. Your friend gives you some more. Now you have 8. How many pennies did your friend give you?

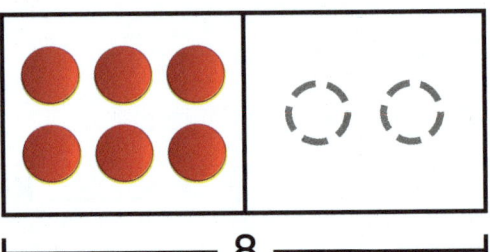

6 plus what is equal to 8?

Addition equation: __6__ + __2__ = __8__

__2__ pennies

Show and Grow I can do it!

1. You have 3 oranges. You buy some more. Now you have 6. How many oranges did you buy?

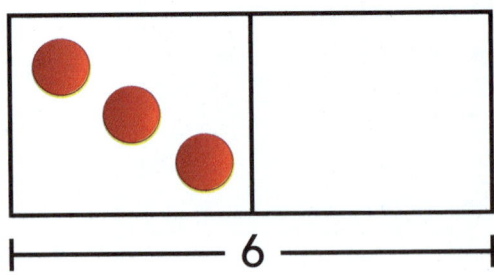

Addition equation: ____ + ____ = ____

____ oranges

46 forty-six

Name _____

✓ Apply and Grow: Practice

2. 5 kids are in a pool. Some more jump in. Now there are 7 kids. How many kids jumped in the pool?

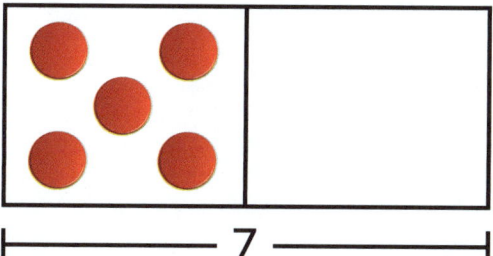

___ + ___ = ___

___ kids

3. You hop 6 times. You hop some more times. You hop 10 times in all. How many more times did you hop?

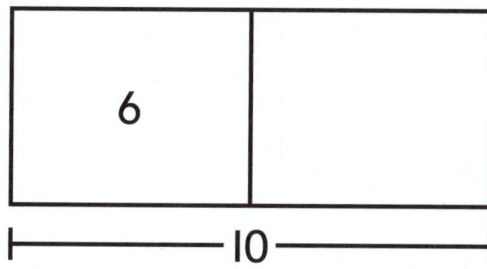

___ + ___ = ___

___ times

4. **Structure** Circle the model that shows the missing number.

2 + ___ = 7

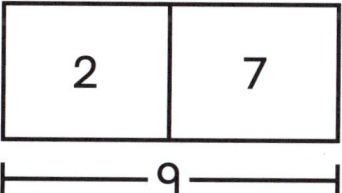

Chapter 1 | Lesson 8

forty-seven 47

Think and Grow: Modeling Real Life

You catch 2 fish before lunch. You catch some more after lunch. You catch 10 in all. How many fish did you catch after lunch?

Model: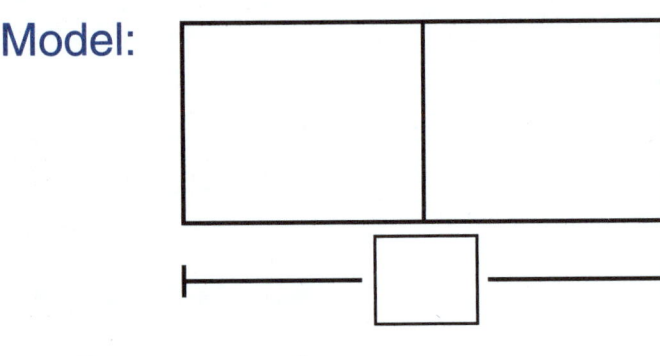

Addition equation:

_____ fish

Show and Grow   I can think deeper!

5. You read 7 books. You read 4 on Monday. You read the rest on Tuesday. How many books do you read on Tuesday?

Model: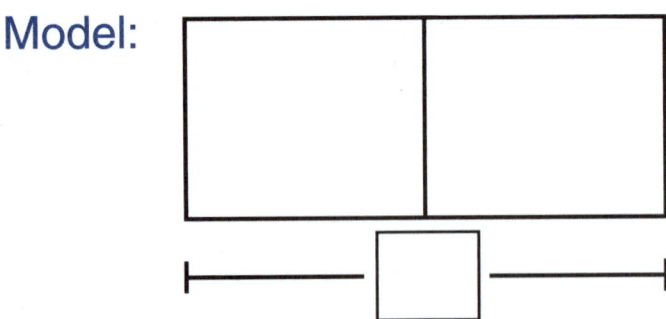

Addition equation:

_____ books

Name _____

Practice 1.8

Learning Target: Solve *add to* word problems that involve a missing addend.

You have 7 stickers. You earn some more. Now you have 9. How many stickers did you earn?

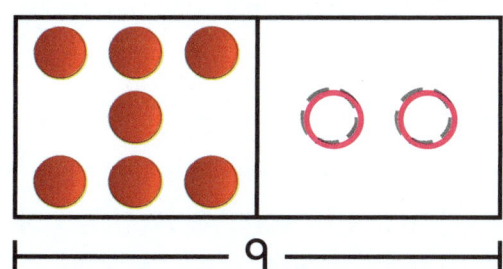

$\underline{7} + \underline{2} = \underline{9}$

$\underline{2}$ stickers

1. 2 kids ride bikes to the park. Some more kids walk to the park. Now there are 6 kids at the park. How many kids walked to the park?

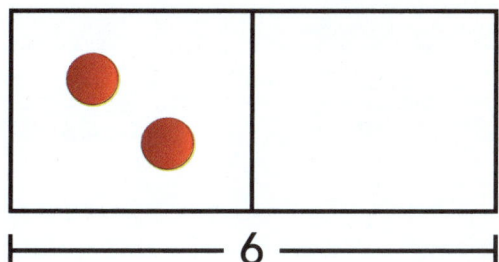

___ + ___ = ___

___ kids

2. You have 5 stamps. You buy some more. Now you have 8. How many stamps did you buy?

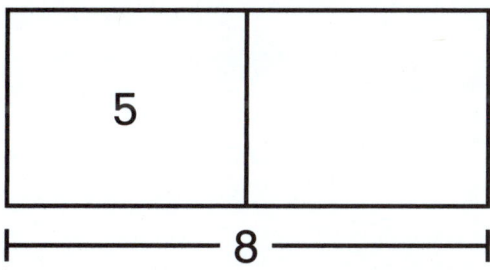

___ + ___ = ___

___ stamps

Chapter 1 | Lesson 8

3. **Structure** Circle the model that shows the missing number.

$$4 + \_\_\_ = 10$$

1	5

—6—

4	2

—6—

4	6

—10—

4. **Modeling Real Life** You tell 7 jokes. You tell some more jokes. You tell 10 jokes in all. How many more jokes did you tell?

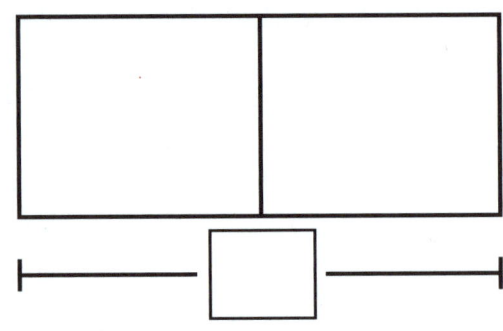

___ jokes

Review & Refresh

5. Write the numbers of bears and honey pots. Are the numbers equal? Circle the thumbs up for *yes* or the thumbs down for *no*.

 ____

 ____

Name _____

Learning Target: Solve word problems that involve putting together and taking apart.

Connect *Put Together* and *Take Apart* Problems

1.9

Explore and Grow

Use counters to model the story.

There are 9 geese. 4 are flying. The rest are on the ground. How many geese are on the ground?

_____ geese

Chapter 1 | Lesson 9

Think and Grow

You have 6 buttons. 4 are green. The rest are black. How many black buttons do you have?

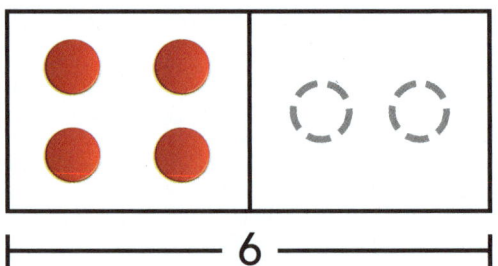

$$4 + 2 = 6$$

$$6 - 4 = 2$$

You can add *or* subtract to find the missing part!

__2__ black buttons

Show and Grow   I can do it!

1. You have 8 beads. 5 are pink. The rest are orange. How many orange beads do you have?

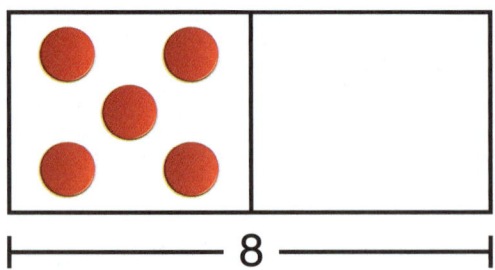

___ + ___ = ___

___ − ___ = ___

___ orange beads

Apply and Grow: Practice

2. You have 9 rings. 7 are green. The rest are orange. How many orange rings do you have?

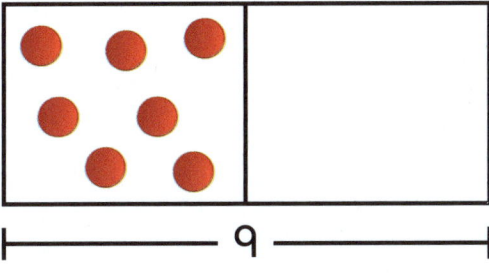

___ + ___ = ___

___ − ___ = ___

___ orange rings

3. There are 3 spiders in a tree. Some more are on the ground. There are 10 spiders in all. How many spiders are on the ground?

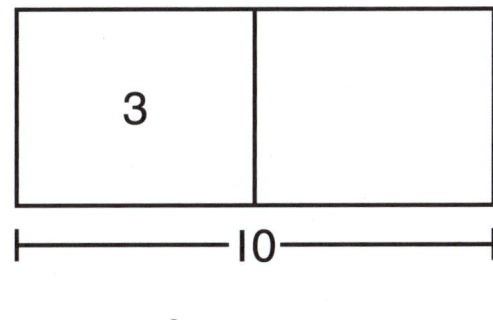

___ + ___ = ___

___ − ___ = ___

___ spiders

4. **Structure** 10 students play instruments. 6 play drums. The rest play flutes. How many students play flutes? Use the model to write an equation and solve.

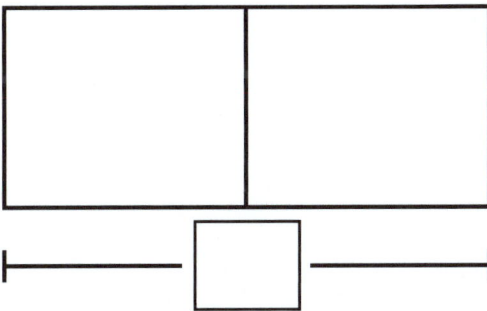

___ ◯ ___ = ___

___ students

Chapter 1 | Lesson 9

 **Think and Grow: Modeling Real Life**

There are 8 students at a playground. 2 are on the slides. The rest are on the swings. How many students are on the swings?

Which equations match the story?

$10 - 2 = 8$ $8 - 2 = 6$

$2 + 6 = 8$ $8 + 2 = 10$

Show how you know:

_____ students are on the swings.

Show and Grow   I can think deeper!

5. Your teacher has 5 outdoor toys. 2 are flying discs. The rest are jump ropes. How many jump ropes are there?

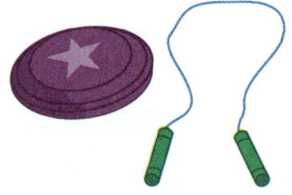

Which equations match the story?

$5 + 2 = 7$ $7 - 2 = 5$

$2 + 3 = 5$ $5 - 2 = 3$

Show how you know:

There are _____ jump ropes.

Name _____

Practice 1.9

Learning Target: Solve word problems that involve putting together and taking apart.

There are 7 cats. 3 are in a window. The rest are on a couch. How many cats are on the couch?

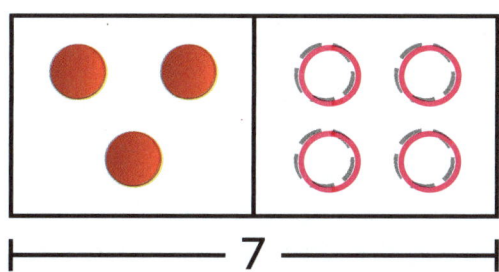

$\underline{3} + \underline{4} = \underline{7}$

$\underline{7} - \underline{3} = \underline{4}$

$\underline{4}$ cats

1. You have 8 markers. 2 are purple. The rest are blue. How many blue markers do you have?

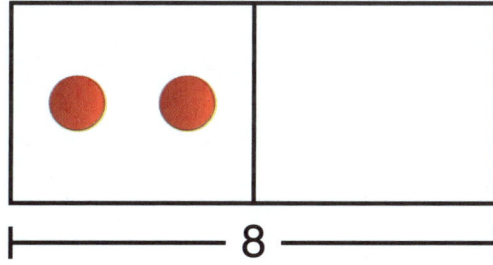

___ + ___ = ___

___ − ___ = ___

___ blue markers

2. There are 6 alligators. 3 are in a swamp. The rest are in the grass. How many alligators are in the grass?

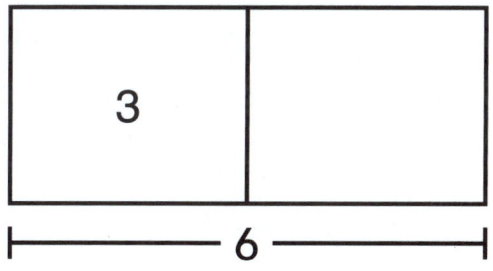

___ + ___ = ___

___ − ___ = ___

___ alligators

Chapter 1 | Lesson 9

fifty-five 55

3. **Structure** There are 2 chickens inside a coop. Some more are outside the coop. There are 10 chickens in all. How many chickens are outside the coop? Use the model to write an equation and solve.

___ ◯ ___ = ___

___ chickens

4. **Modeling Real Life** Your teacher has 7 balls. 2 are soccer balls. The rest are basketballs. How many basketballs are there?

Which equations match the story?

$7 - 2 = 5$ $9 - 2 = 7$ $7 + 2 = 9$ $2 + 5 = 7$

Show how you know:

There are ___ basketballs.

Review & Refresh

Find the sum.

5. $6 + 0 = $ ___

6. $0 + 0 = $ ___

7. $5 + 1 = $ ___

8. $9 + 1 = $ ___

Name _____

Performance Task 1

1. 3 birds are on a wire. 5 birds are on a branch.

   a. 2 more birds land on the wire. How many birds are on the wire now?

   _____ birds

   b. 1 bird flies away from the branch. How many birds are on the branch now?

   _____ birds

   c. How many more birds are on the wire now than on the branch now?

   _____ more birds

2. There are 6 birds. Draw some on the wire and some on the branch. Write an addition equation and a subtraction equation to match your picture.

   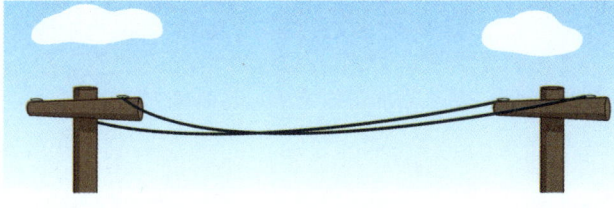 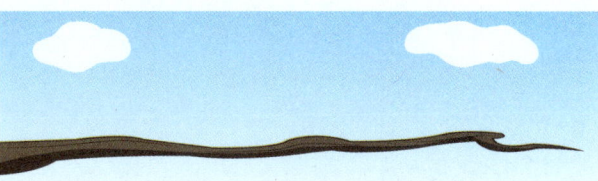

   ___ ◯ ___ = ___   ___ ◯ ___ = ___

Three in a Row

To Play: Place the Three in a Row Game Cards in a pile. Players take turns. On your turn, flip over the top card and solve the problem. Place a counter on the answer. Your turn is over. Repeat until a player gets three in a row.

Name _____

Chapter Practice 1

1.1 Addition: Add To

1.

   There are 4 🐸.

   2 more 🐸 join them.

   Now there are ____ 🐸.

   ___ + ___ = ___

1.2 Solve Add To Problems

2. There are 5 🐙.

   3 more 🐙 join them.

   How many 🐙 are there now?

   ___ + ___ = ___

   ___ 🐙

3. **Modeling Real Life** You have 3 🖍. Your friend gives you 1 🖍. How many 🖍 do you have now?

   ___ 🖍

1.3 Solve *Put Together* Problems

4. You have 4 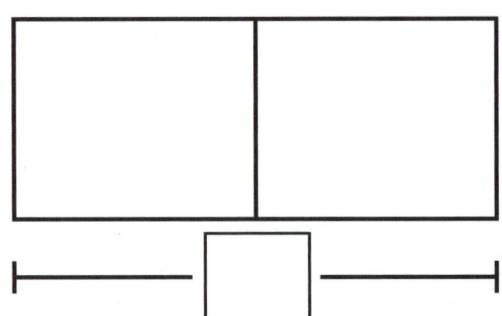 and 4 . How many flowers do you have in all?

___ + ___ = ___

___ flowers

1.4 Solve *Put Together* Problems with Both Addends Unknown

5. There are 10 . Some are standing up. The rest are knocked down. Draw two different pictures to show the .

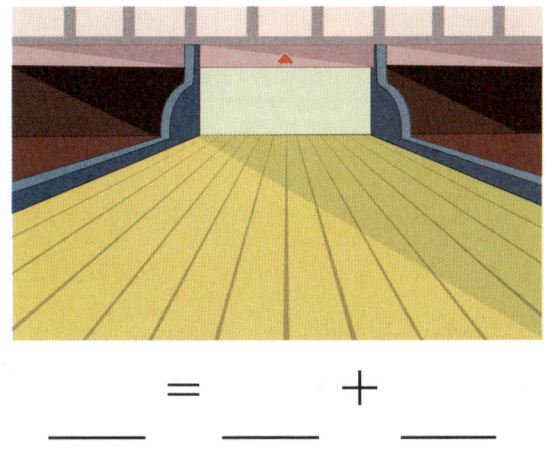

___ = ___ + ___

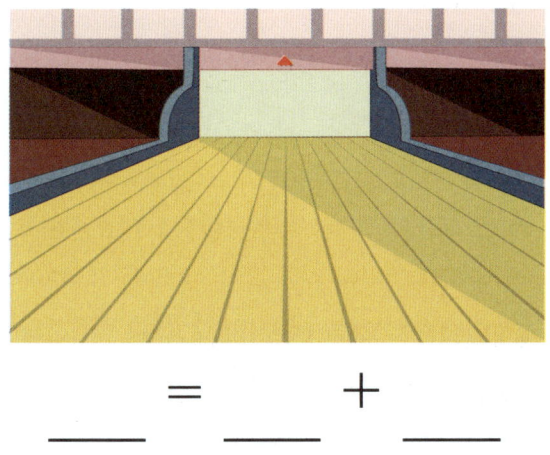

___ = ___ + ___

1.5 Solve *Take From* Problems

6. There are 7 .

3 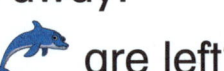 swim away.

How many  are left?

___ − ___ = ___

___

60   sixty

1.6 Solve *Compare* Problems: More

7. You have 6 whistles. Your friend has 3 whistles. How many more whistles do you have?

   ____ − ____ = ____

____ more whistles

8. There are 2 yellow blankets and 9 green blankets. Draw the missing blankets. How many more green blankets are there?

____ − ____ = ____

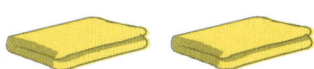

____ more green blankets

1.7 Solve *Compare* Problems: Fewer

9. You have 5 puzzles. Your friend has 7 puzzles. How many fewer puzzles do you have?

   ____ − ____ = ____

____ fewer puzzles

10. There are 8 red leaves and 4 orange leaves. Draw the missing leaves. How many fewer orange leaves are there?

   ____ − ____ = ____

____ fewer orange leaves

 **Solve *Add To* Problems with Change Unknown**

11. You have 3 bracelets. Your friend gives you some more. Now you have 9. How many bracelets did your friend give you?

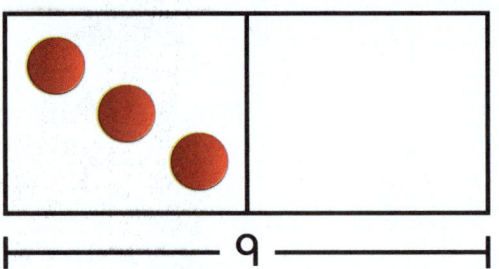

___ + ___ = ___

___ bracelets

12. You have 5 carrots. You buy some more. Now you have 10. How many carrots did you buy?

___ + ___ = ___

___ carrots

 **Connect *Put Together* and *Take Apart* Problems**

13. You have 4 hats. 2 have polka dots. The rest have stripes. How many striped hats do you have?

___ + ___ = ___

___ − ___ = ___

___ striped hats

62   sixty-two

2 Fluency and Strategies within 10

- What is your favorite kind of flower?
- How many flowers do you see? If you pick 5 flowers, how many flowers will be left?

Chapter Learning Target:
Understand fluency and strategies.

Chapter Success Criteria:
- I can identify strategies.
- I can describe equations.
- I can explain rules.
- I can apply strategies.

Name _____

2 Vocabulary

Organize It

Review Words
addend
sum
difference

Use the review words to complete the graphic organizer.

[   ]
↑   ↑
3 + 4 = 7
↓
[   ]

5 − 1 = 4
↓
[   ]

Define It

Use your vocabulary cards to match.

1. number line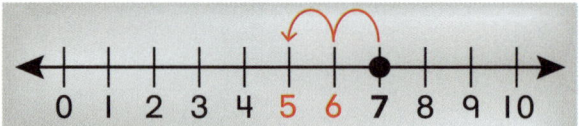

2. count on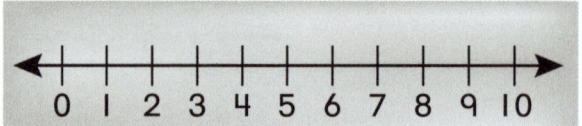

3. count back

64 sixty-four

Chapter 2 Vocabulary Cards

count back	count on
doubles	doubles minus 1
doubles plus 1	number line

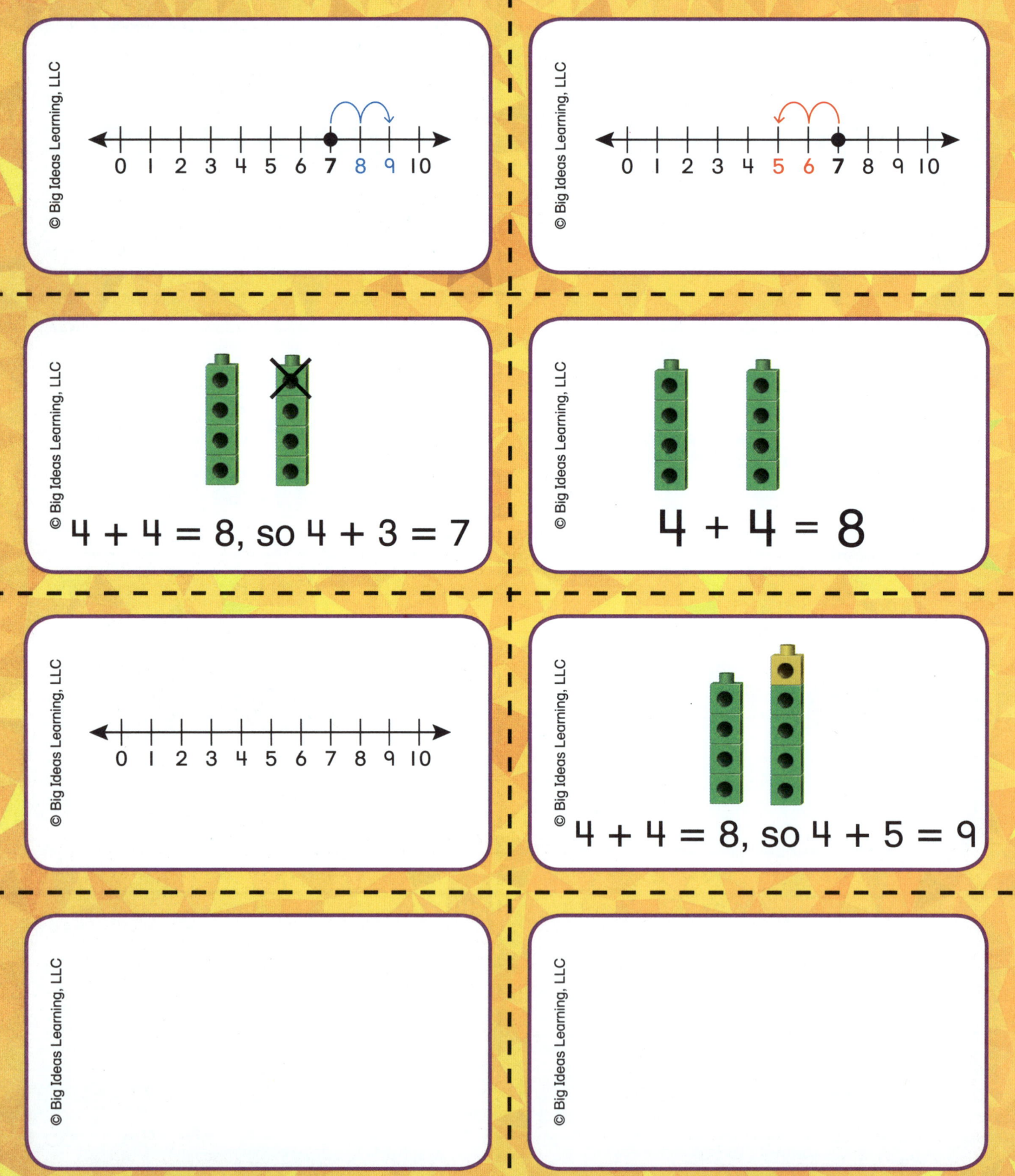

Name _____

Learning Target: Solve equations when an addend is 0.

 Add 0 **2.1**

Explore and Grow

Use linking cubes to model each story.

There are 6 ducks in the pond. 0 ducks join them. How many ducks are in the pond now?

_____ ducks

There are 0 ducks in the pond. 8 ducks go in the pond. How many ducks are in the pond now?

_____ ducks

Chapter 2 | Lesson 1

sixty-five 65

Think and Grow

"When you add 0 to a number, the sum is that number."

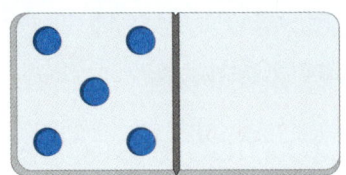

5 + 0 = 5

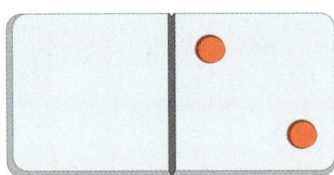

0 + 2 = 2

"When you add a number to 0, the sum is that number."

Show and Grow — I can do it!

Use the picture to write an equation.

1.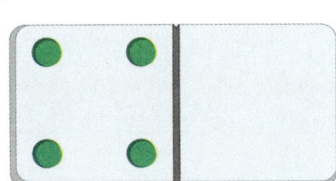

___ + 0 = ___

2.

0 + ___ = ___

3.

0 + ___ = ___

4.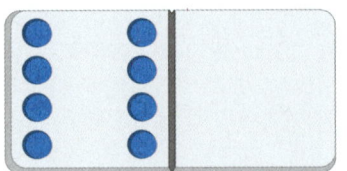

___ + 0 = ___

Name _____

✓ Apply and Grow: Practice

Use the picture to write an equation.

5.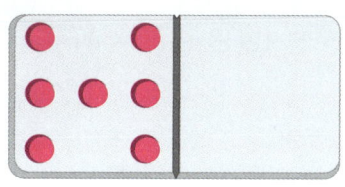

___ + 0 = ___

6.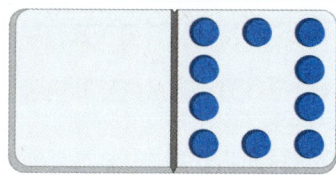

0 + ___ = ___

7. 9 + 0 = ___

8. 0 + 8 = ___

DIG DEEPER! Find each sum. Think: What do you notice?

9. 3 + 0 = ___

   0 + 3 = ___

10. 0 + 6 = ___

    6 + 0 = ___

11. **Logic** There are 7 penguins in all. How many penguins are inside the igloo?

___ penguins

Chapter 2 | Lesson 1

Think and Grow: Modeling Real Life

There are no students at a bus stop. Then 9 students arrive. How many students are at the bus stop now?

Model:

Addition equation: _____

_____ students

Show and Grow — I can think deeper!

12. Your friend does not have any tokens. You give your friend 7 tokens. How many tokens does your friend have now?

Model: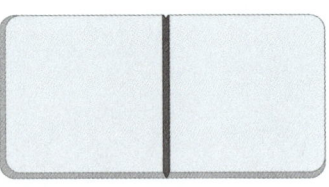

Addition equation: _____

_____ tokens

Name _____

Practice 2.1

Learning Target: Solve equations when an addend is 0.

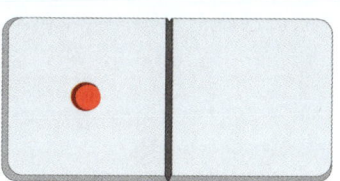 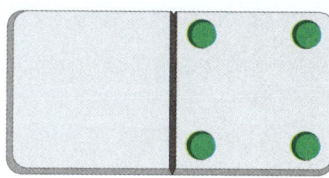

__1__ + 0 = __1__   0 + __4__ = __4__

Use the picture to write an equation.

1.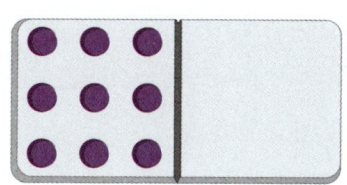

   ___ + 0 = ___

2.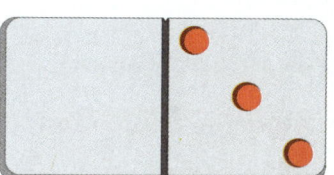

   0 + ___ = ___

3. 10 + 0 = ___

4. 0 + 5 = ___

5. 0 + 6 = ___

6. 0 + 0 = ___

Chapter 2 | Lesson 1

sixty-nine 69

7. **DIG DEEPER!** Find each sum. Think: What do you notice?

    7 + 0 = ___

    0 + 7 = ___

8. **Logic** There are 8 students in all. How many students are inside the museum?

    ____ students

9. **Modeling Real Life** There are no seals on the shore. Then 10 seals swim to the shore. How many seals are on the shore now?

    ____ seals

Review & Refresh

Write the number of goldfish.

10.
    ____

11.
    ____

12.
    ____

13.
    ____

Name _____

Learning Target: Subtract 0 and subtract all.

Subtract 0 and Subtract All 2.2

Explore and Grow

Use linking cubes to model each story.

There are 5 beavers on the log. None of the beavers leave. How many beavers are left?

____ beavers

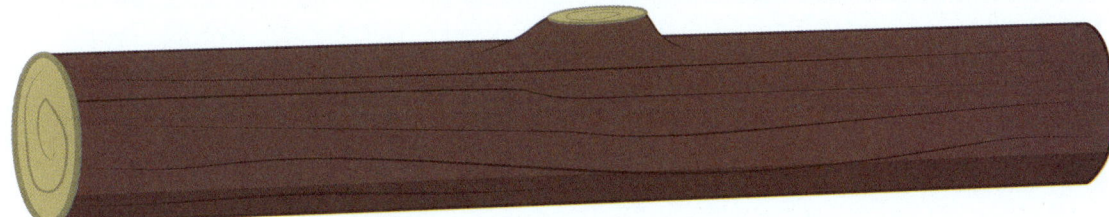

There are 5 beavers on the log. All of the beavers leave. How many beavers are left?

____ beavers

Chapter 2 | Lesson 2

 **Think and Grow**

When you subtract 0 from a number, the difference is that number.

$\underline{3} - 0 = \underline{3}$

When you subtract a number from itself, the difference is 0.

$\underline{3} - \underline{3} = 0$

Show and Grow — I can do it!

Use the picture to write an equation.

1.

   $\underline{} - 0 = \underline{}$

2.

   $\underline{} - \underline{} = 0$

3.

   $\underline{} - 0 = \underline{}$

4.

   $\underline{} - \underline{} = 0$

Name _____

✓ Apply and Grow: Practice

Use the picture to write an equation.

5.

___ − 0 = ___

6.

___ − ___ = 0

7. 5 − 5 = ___

8. 6 − 0 = ___

9. 9 − 0 = ___

10. 7 − 7 = ___

11. **MP Structure** Complete the equation. Then use the words to complete the sentence.

4 − ___ = 4

When you _____ 0 from a number,

the _____ is that _____.

Words
difference
number
subtract

Chapter 2 | Lesson 2

# Think and Grow: Modeling Real Life

Your friend has 7 pennies. You have 7 fewer pennies than your friend. How many do you have?

Draw a picture:

Subtraction equation:

_____ pennies

Show and Grow   I can think deeper!

12. You have 6 pieces of chalk. You give all of your chalk to your friend. How many pieces do you have left?

    Draw a picture:

    Subtraction equation:

    _____ pieces of chalk

Name _____

Practice 2.2

Learning Target: Subtract 0 and subtract all.

_6_ – 0 = _6_   _6_ – _6_ = 0

Use the picture to write an equation.

1.

___ – 0 = ___

2.

___ – ___ = 0

3. 3 – 3 = ___

4. 1 – 0 = ___

5. 1 – 1 = ___

6. 7 – 0 = ___

Chapter 2 | Lesson 2

7. **Structure** Complete the equation. Then use the words to complete the sentence.

9 − ____ = 0

When you _____ a _____ from itself, the _____ is 0.

Words
difference
number
subtract

8. Modeling Real Life You have 4 stuffed animals. You give all of them to your friend. How many stuffed animals do you have left?

____ stuffed animals

Review & Refresh

Write the number of dots you see on each domino. Then write the numbers in order.

9.

____ ____ ____ 3

10.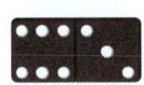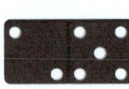

____ ____ ____ ____

Name _____

Learning Target: Add and subtract 1.

Explore and Grow

Use linking cubes to model each story.

There are 7 kids on the bench. 1 kid joins them. How many kids are on the bench now?

_____ kids

There are 8 kids on the bench. 1 kid leaves. How many kids are left?

_____ kids

Chapter 2 | Lesson 3

Think and Grow

When you add 1 to a number, the sum is the next number when counting.

4 + 1 = 5

When you subtract 1 from a number, the difference is the previous number when counting.

4 − 1 = 3

Show and Grow I can do it!

Use the picture to write an equation.

1.

___ − ___ = ___

2.

___ + ___ = ___

Name _____

✓ Apply and Grow: Practice

Use the picture to write an equation.

3.

___ + ___ = ___

4.

___ − ___ = ___

5. 5 − 1 = ___

6. 6 + 1 = ___

7. 8 + 1 = ___

8. 7 − 1 = ___

DIG DEEPER! Circle the problem with the greater sum or difference.

9.   2 − 1   2 + 1

10.   5 + 1   6 − 1

11. **YOU BE THE TEACHER** Circle to show who is correct. Show how you know.

3 + 1 = 2

3 + 1 = 4

Chapter 2 | Lesson 3

seventy-nine 79

Think and Grow: Modeling Real Life

You have 9 action figures. Newton has 1 more than you. Descartes has 1 fewer than you. Who has more, Newton or Descartes?

Equations: Newton   Descartes

Who has more?   Newton   Descartes

Show and Grow   I can think deeper!

12. You have 5 video games. Newton has 1 fewer than you. Descartes has 1 more than you. Who has fewer, Newton or Descartes?

    Equations: Newton   Descartes

    Who has fewer?   Newton   Descartes

Name _____

Practice 2.3

Learning Target: Add and subtract 1.

2 + 1 = 3

2 − 1 = 1

Use the picture to write an equation.

1.

___ + ___ = ___

2.

___ − ___ = ___

3. 10 − 1 = ___

4. 7 + 1 = ___

DIG DEEPER! Circle the problem with the greater sum or difference.

5.   4 − 1   4 + 1 | 6.   7 + 1   8 − 1

---

7. **YOU BE THE TEACHER** Circle to show who is correct. Show how you know.

   6 − 1 = __5__   6 − 1 = __7__

---

8. **Modeling Real Life** You have 3 karate belts. Newton has 1 fewer than you. Descartes has 1 more than you. Who has fewer, Newton or Descartes?

   Who has fewer?   Newton   Descartes

Review & Refresh

9. Use the picture to complete the number bond.

 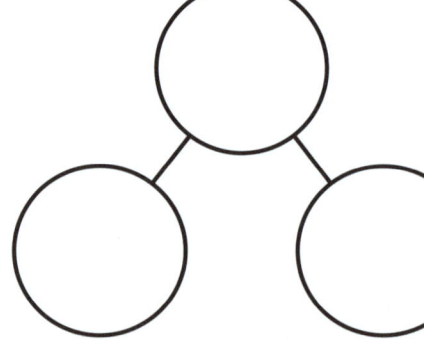

Name _____

Add Doubles from 1 to 5
2.4

Learning Target: Find the sum of doubles from 1 to 5.

Explore and Grow

Use counters to model the story.

You have 3 balls. Your friend has 3 balls. How many balls are there in all?

_____ balls

Chapter 2 | Lesson 4

 **Think and Grow**

The addends are the same.

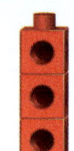 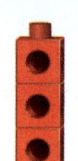

3 + 3 = 6

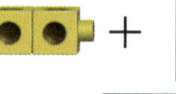

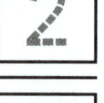 +

2
2

4

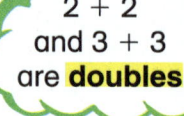

2 + 2 and 3 + 3 are **doubles**.

Show and Grow — I can do it!

1.

___ + ___ = ___

2.

___ + ___ = ___

3.

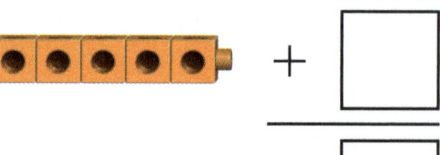

4.

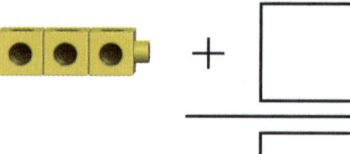

84 eighty-four

Name _____

Apply and Grow: Practice

5.

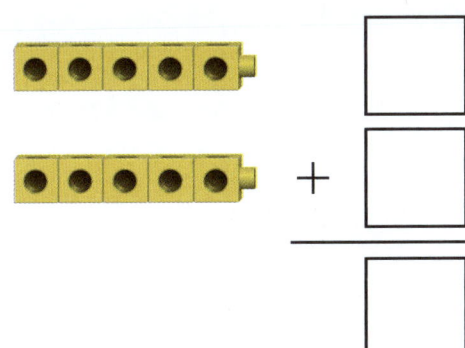

6.

___ + ___ = ___

7. 2 + 2 = ___

8. 5 + 5 = ___

9.
```
    3
+   3
  ___
```

10.
```
    4
+   4
  ___
```

11. **DIG DEEPER!** Circle the equations you can complete using doubles.

___ + ___ = 6   ___ + ___ = 5

___ + ___ = 9   ___ + ___ = 2

Chapter 2 | Lesson 4

# Think and Grow: Modeling Real Life

You and your friend color the same number of pictures. There are 10 pictures in all. How many pictures do you each color?

Draw a picture:

Addition equation:

_____ pictures

Show and Grow   *I can think deeper!*

12. You and your friend have the same number of flowers. There are 8 flowers in all. How many flowers do you each have?

    Draw a picture:

    Addition equation:

    _____ flowers

Name _____

Practice 2.4

Learning Target: Find the sum of doubles from 1 to 5.

$\underline{1} + \underline{1} = \underline{2}$

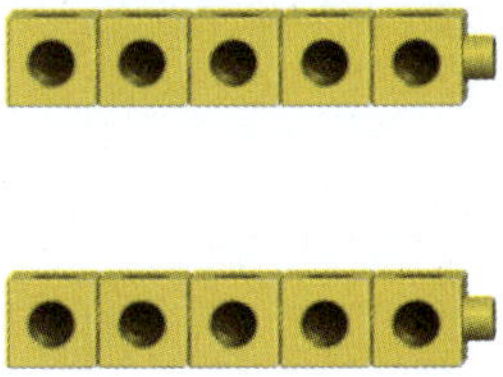

$\begin{array}{r} 5 \\ + 5 \\ \hline 10 \end{array}$

1. ___ + ___ = ___

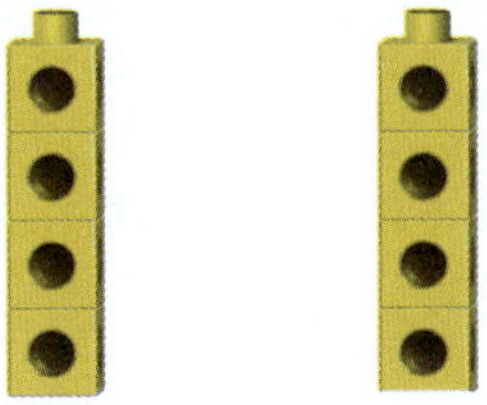

2. ___ + ___ = ___

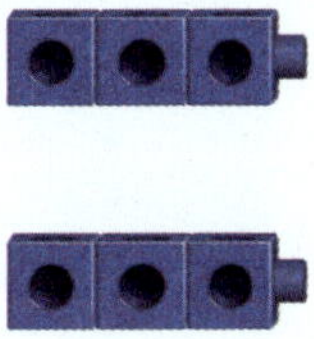

3. $2 + 2 = $ ___

4. $4 + 4 = $ ___

5. $\begin{array}{r} 5 \\ + 5 \\ \hline \end{array}$

6. $\begin{array}{r} 3 \\ + 3 \\ \hline \end{array}$

Chapter 2 | Lesson 4

7. **DIG DEEPER!** Circle the equations you can complete using doubles.

___ + ___ = 3   ___ + ___ = 8

___ + ___ = 4   ___ + ___ = 7

8. **Modeling Real Life** Newton and Descartes each have the same number of linking cubes. There are 6 linking cubes in all. How many linking cubes do Newton and Descartes each have?

___ linking cubes

Review & Refresh

Use the picture to write an equation.

9.

___ + ___ = ___

10.

___ + ___ = ___

Name _____

Learning Target: Use the *doubles plus 1* and *doubles minus 1* strategies to find a sum.

Explore and Grow

Use counters to model the story.

You collect 4 shells. Your friend collects 4 shells. How many shells are there in all?

_____ shells

You collect 4 shells. Your friend collects 5 shells. How many shells are there in all?

_____ shells

Chapter 2 | Lesson 5

 **Think and Grow**

Use the double 4 + 4 to find each sum.

4 + 5 = __9__   4 + 3 = __7__

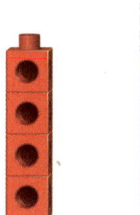   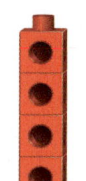 

doubles plus 1   **doubles minus 1**

4 + 5 is equal to 4 + 4 and 1 more.   4 + 3 is equal to 1 less than 4 + 4.

Show and Grow — I can do it!

Use the double 3 + 3 to find each sum.

1. 3 + 4 = ___

     3
           + 2
           ___

Name _____

✓ Apply and Grow: Practice

Use the double 2 + 2 to find each sum.

2. 2 + 3 = ____

2 + 1 = ____

Find the sum. Write the double you used.

3. 3 + 4 = ____

____ + ____ = ____

4.

```
    5
+   4
─────
  ☐
```

```
    ☐
+   ☐
─────
  ☐
```

5. **Number Sense** Use each card once to write two addition equations.

3   2   2
2   4   5

____ + ____ = ____

____ + ____ = ____

Chapter 2 | Lesson 5

ninety-one 91

# Think and Grow: Modeling Real Life

You eat 4 grapes. Your friend eats 1 more than you. How many grapes do you and your friend eat in all?

Which doubles can you use to find the sum?

4 + 4   5 + 5   3 + 3

Addition equation: _____

_____ grapes

Show and Grow   I can think deeper!

6. You have 5 toy cars. Your friend has 1 fewer than you. How many cars do you and your friend have in all?

Which doubles can you use to find the sum?

4 + 4   6 + 6   5 + 5

Addition equation: _____

_____ toy cars

Name _____

Practice 2.5

Learning Target: Use the *doubles plus 1* and *doubles minus 1* strategies to find a sum.

Use the double 3 + 3 to find each sum.

3 + 4 = __7__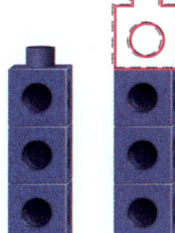

3 + 2 = __5__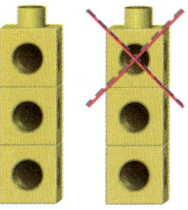

Use the double 4 + 4 to find each sum.

1. 4 + 5 = ___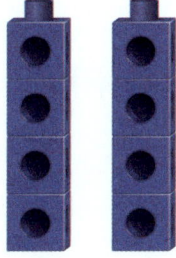

4 + 3 = ___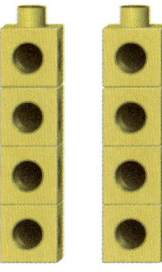

Find the sum. Write the double you used.

2. 1 + 2 = ___

___ + ___ = ___

3. 3 + 2 = ___

___ + ___ = ___

Chapter 2 | Lesson 5

4. **Number Sense** Use each card once to write two addition equations.

3   6   2
  5   3   3

___ + ___ = ___

___ + ___ = ___

5. **Modeling Real Life** Newton catches 2 butterflies. Descartes catches 1 more than Newton. How many butterflies do Newton and Descartes catch in all?

Which doubles can you use to find the sum?

3 + 3   1 + 1   2 + 2

_____ butterflies

Review & Refresh

6. Circle the model that shows the missing number.

2 + ___ = 5

2	5		2	3		3	5
---	---		---	---		---	---
7		5		8

Name _____

Add in Any Order 2.6

Learning Target: Add in any order to find a sum.

Explore and Grow

Use counters to model each problem. What do you notice?

$$4 + 3 = \underline{}$$

$$3 + 4 = \underline{}$$

Chapter 2 | Lesson 6

 Think and Grow

5 + 3 = 8   3 + 5 = 8

Change the order of the addends.   The sum stays the same.

Show and Grow   I can do it!

1. ___ + ___ = ___   ___ + ___ = ___

2. ___ + ___ = ___   ___ + ___ = ___

3.

Name _____

Apply and Grow: Practice

4.

___ + ___ = ___   ___ + ___ = ___

Find the sum. Then change the order of the addends. Write the new addition problem.

5.
```
   1          ▢
 + 4       + ▢
 ___        ___
   ▢          ▢
```

6.
```
   5          ▢
 + 2       + ▢
 ___        ___
   ▢          ▢
```

7. 2 + 6 = ___

___ + ___ = ___

8. ___ = 8 + 1

___ = ___ + ___

9. **Number Sense** Use the numbers shown to write two addition equations.

4   9   5

___ + ___ = ___   ___ + ___ = ___

Chapter 2 | Lesson 6

ninety-seven 97

Think and Grow: Modeling Real Life

You have 7 shirts. 3 are green. The rest are blue. How many blue shirts do you have?

Which equations describe your shirts?

$3 + 4 = 7$ $3 + 7 = 10$

$7 + 3 = 10$ $4 + 3 = 7$

Show how you know:

You have ____ blue shirts.

Show and Grow   I can think deeper!

10. You have 5 cups. 2 are yellow. The rest are red. How many red cups do you have?

    Which equations describe your cups?

    $5 + 2 = 7$ $2 + 3 = 5$

    $2 + 5 = 7$ $3 + 2 = 5$

    Show how you know:

    You have ____ red cups.

Name _____

Practice 2.6

Learning Target: Add in any order to find a sum.

```
    3              2
  + 2            + 3
  ───            ───
    5              5
```

1.

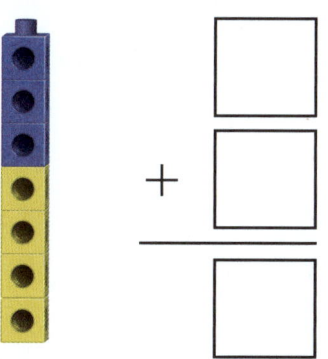

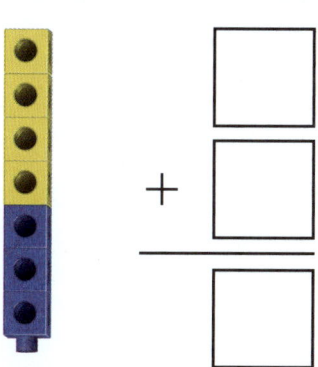

2.

___ + ___ = ___   ___ + ___ = ___

Find the sum. Then change the order of the addends.
Write the new addition problem.

3.
```
    4
  + 6
  ───
```

4.

___ = 5 + 1

___ = ___ + ___

Chapter 2 | Lesson 6

ninety-nine 99

5. **Number Sense** Use the numbers shown to write two addition equations.

8   10   2

___ + ___ = ___   ___ + ___ = ___

6. **Modeling Real Life** You have 7 tomatoes. 2 are red. The rest are yellow. How many yellow tomatoes do you have?

Which equations describe your tomatoes?

$2 + 5 = 7$ $5 + 2 = 7$

$7 + 2 = 9$ $2 + 7 = 9$

Show how you know:

You have ___ yellow tomatoes.

Review & Refresh

Use the ten frame to complete the equation.

7.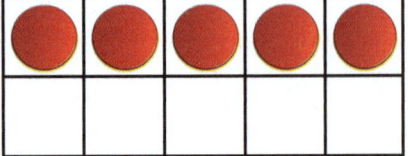

8.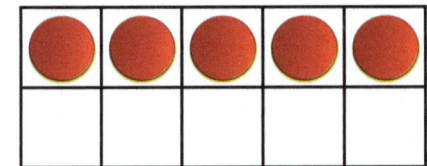

$5 +$ ___ $= 8$ $5 +$ ___ $= 10$

Name _____

Count On to Add

Learning Target: Use the *count on* strategy to find a sum.

Model the story.

There are 5 coins in a piggy bank. You put in 2 more. How many coins are in the bank now?

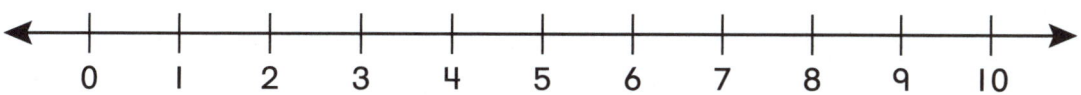

_____ coins

Chapter 2 | Lesson 7

 **Think and Grow**

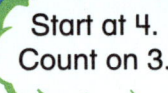 Start at 4. Count on 3.

4 + 3 = 7

To add, **count on**.

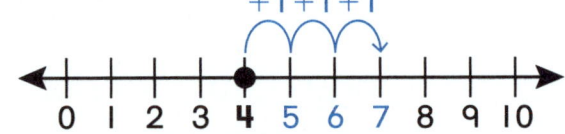

This **number line** shows the numbers 0 through 10.

Show and Grow I can do it!

1. 7 + 2 = ___

2. 3 + 1 = ___

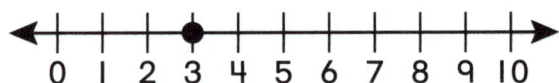

3. 4 + 6 = ___

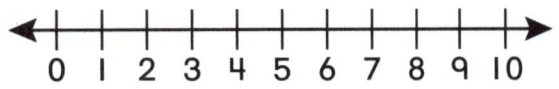

4. 0 + 5 = ___

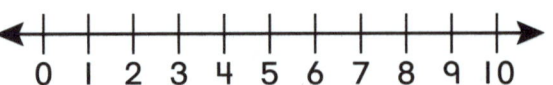

Name _____

 **Apply and Grow: Practice**

5. $4 + 2 =$ _____

6. $6 + 1 =$ _____

7. 5
   $+4$
   _____
   ☐

8. 3
   $+2$
   _____
   ☐

9. _____ $= 0 + 7$

10. _____ $= 2 + 8$

11. **DIG DEEPER!** Tell what problems Newton and Descartes solved. Think: How are the problems the same? How are they different?

 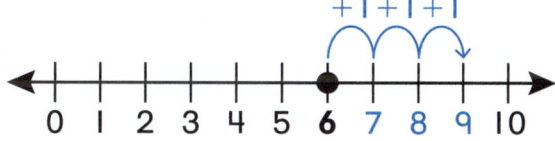   ___ + ___ = ___

 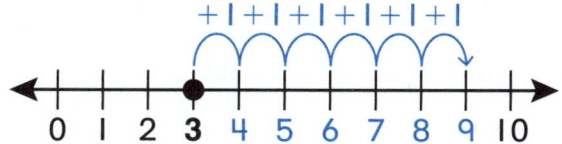   ___ + ___ = ___

Chapter 2 | Lesson 7   one hundred three   103

 Think and Grow: Modeling Real Life

You and your friend are on a scavenger hunt. You find 3 clues. You and your friend find 8 clues in all. How many clues does your friend find?

Model:

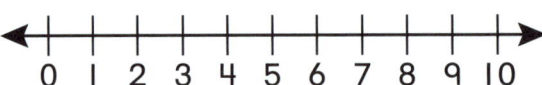

_____ clues

Show and Grow   I can think deeper!

12. Your friend collects 4 cans. You and your friend collect 10 cans in all. How many cans do you collect?

Model:

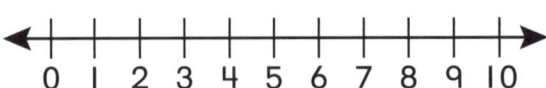

_____ cans

104   one hundred four

Name _____

Practice 2.7

Learning Target: Use the *count on* strategy to find a sum.

$8 + 2 = \underline{10}$

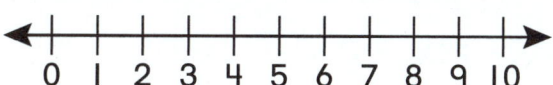

1. $4 + 3 = \underline{}$

2. $8 + 1 = \underline{}$

3. $6 + 2 = \underline{}$

4. $7 + 3 = \underline{}$

5. $\begin{array}{r} 0 \\ +9 \\ \hline \square \end{array}$

6. $\underline{} = 1 + 5$

Chapter 2 | Lesson 7

one hundred five

7. **DIG DEEPER!** Tell what problems Newton and Descartes solved. Think: How are the problems the same? How are they different?

 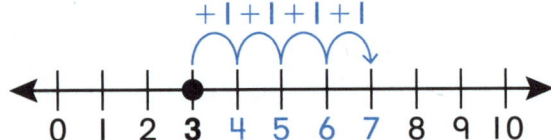   ___ + ___ = ___

 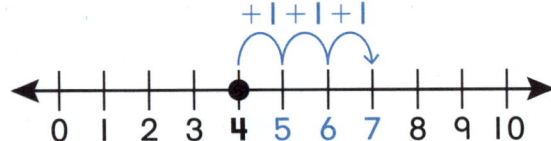   ___ + ___ = ___

8. **Modeling Real Life** You have 4 coins. You and your friend have 9 coins in all. How many coins does your friend have?

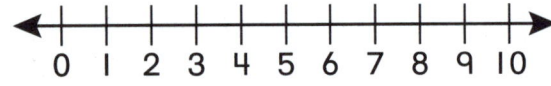

___ coins

Review & Refresh

Use the picture to write an equation.

9.

___ ___ = ___

10.

___ ___ = ___

106 one hundred six

Name _____

Count Back to Subtract 2.8

Learning Target: Use the *count back* strategy to find a difference.

Explore and Grow

Model the story.

There are 8 students in a line. 2 of them leave. How many students are left?

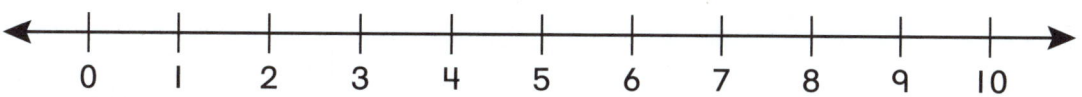

_____ students

Chapter 2 | Lesson 8

 Think and Grow

To subtract, **count back**.

8 − 3 = 5

Start at 8. Count back 3.

Show and Grow *I can do it!*

1. 5 − 4 = ___

2. 7 − 3 = ___

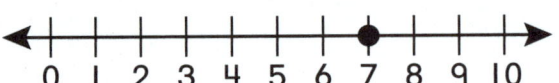

3. 6 − 1 = ___

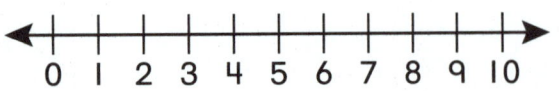

4. 10 − 8 = ___

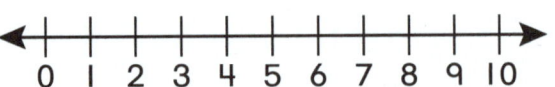

108   one hundred eight

Name ______________________________

Apply and Grow: Practice

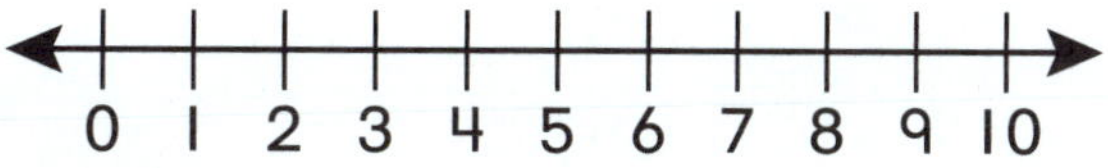

5. $10 - 6 =$ ____

6. $9 - 3 =$ ____

7.

$$\begin{array}{r} 5 \\ -\ 2 \\ \hline \square \end{array}$$

8.

$$\begin{array}{r} 8 \\ -\ 0 \\ \hline \square \end{array}$$

9. ____ $= 7 - 6$

10. ____ $= 4 - 1$

11. MP **Structure** Write the problem shown.

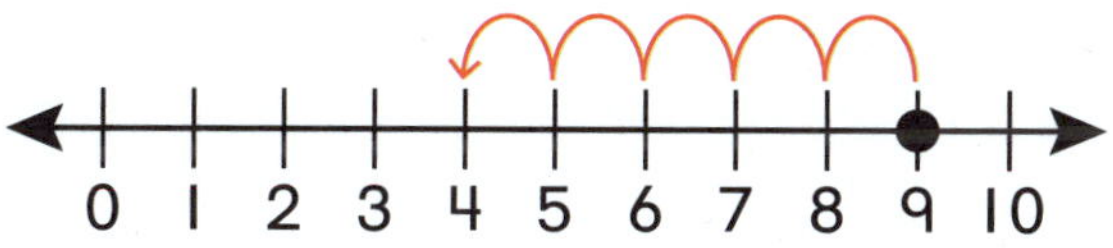

____ − ____ = ____

 **Think and Grow: Modeling Real Life**

There are 5 students on one side of a table and 5 students on the other. 7 students leave. How many students are left?

Model:

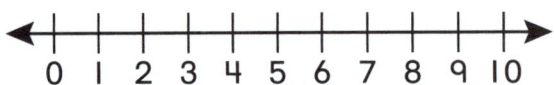

_____ students

Show and Grow   I can think deeper!

12. You have 4 board games and 4 card games. You give 3 games to your friend. How many games do you have left?

    Model:

    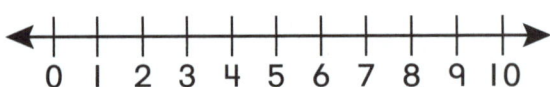

    _____ games

110   one hundred ten

Name _____

Practice 2.8

Learning Target: Use the *count back* strategy to find a difference.

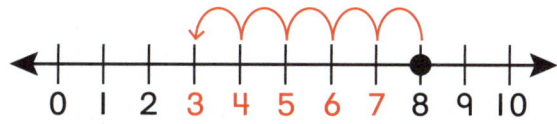

8 − 5 = __3__

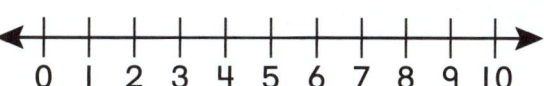

1. 5 − 4 = ___

2. 8 − 3 = ___

3. 9 − 1 = ___

4. 6 − 2 = ___

5. 7
  − 5
  ____
  ☐

6. ___ = 10 − 5

Chapter 2 | Lesson 8   one hundred eleven   111

7. **Structure** Write the problem shown.

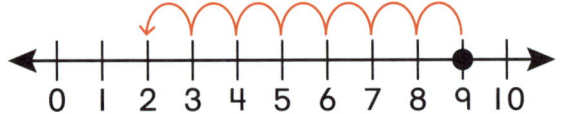

____ − ____ = ____

8. Modeling Real Life You have 4 star stickers and 4 heart stickers. You give 6 away. How many stickers do you have left?

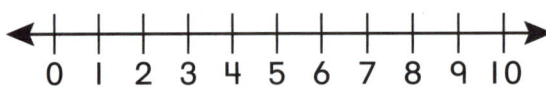

____ stickers

Review & Refresh

9. Use the pictures to write the related equations.

____ ○ ____ = ____   ____ ○ ____ = ____

Name _____

Use Addition to Subtract 2.9

Learning Target: Use the *add to subtract* strategy to find a difference.

Explore and Grow

Use counters to model each problem.

$$4 + \underline{} = 7$$

$$7 - 4 = \underline{}$$

Chapter 2 | Lesson 9

one hundred thirteen 113

Think and Grow

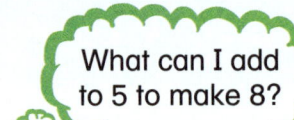

8 − 5 = ?

Count on 3 to make 8.

Think 5 + __3__ = 8. So, 8 − 5 = __3__.

Show and Grow — I can do it!

1. 5 − 4 = ?

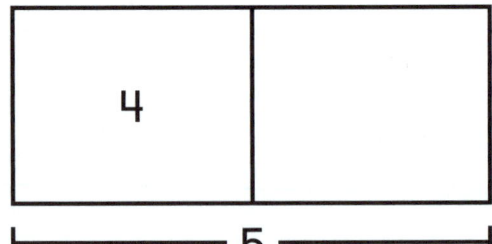

Think 4 + ___ = 5.

So, 5 − 4 = ___.

2. 6 − 3 = ?

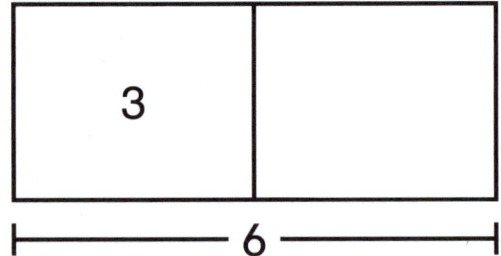

Think 3 + ___ = 6.

So, 6 − 3 = ___.

114   one hundred fourteen

Name _____

 Apply and Grow: Practice

3. $8 - 4 = ?$

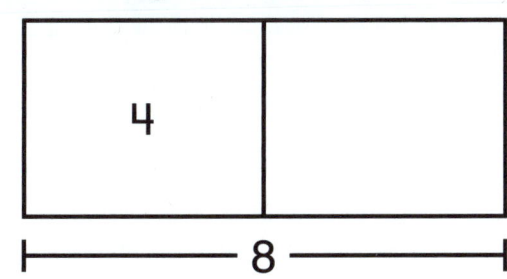

Think $4 +$ ____ $= 8$.

So, $8 - 4 =$ ____.

4. $7 - 5 = ?$

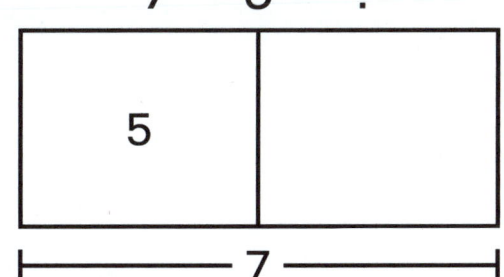

Think $5 +$ ____ $= 7$.

So, $7 - 5 =$ ____.

5. $6 - 4 = ?$

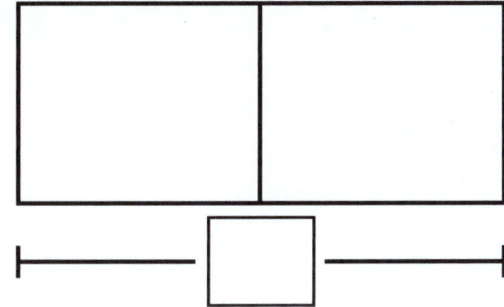

Think ____ $+$ ____ $=$ ____.

So, $6 - 4 =$ ____.

6. $9 - 6 = ?$

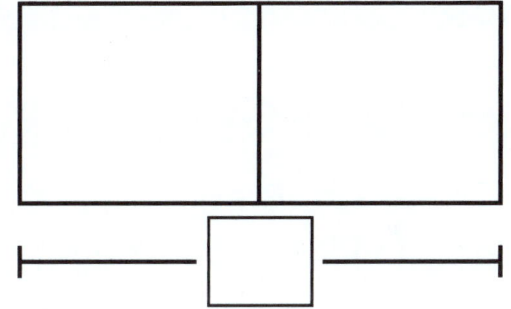

Think ____ $+$ ____ $=$ ____.

So, $9 - 6 =$ ____.

7. **YOU BE THE TEACHER** There are 8 goats. 2 of them leave. Newton uses addition to tell how many goats are left. Is he correct? Show how you know.

$8 + 2 = 10$
10 goats are left.

Chapter 2 | Lesson 9

one hundred fifteen 115

Think and Grow: Modeling Real Life

There are 10 puppies. 7 are brown. The rest are yellow. How many puppies are yellow?

Model:

Subtraction equation:

_____ puppies

Show and Grow — I can think deeper!

8. There are 8 eggs. 1 of them hatches. How many eggs still need to hatch?

Model:

Subtraction equation:

_____ eggs

Name _____

Practice 2.9

Learning Target: Use the *add to subtract* strategy to find a difference.

5 − 4 = ?

[4 | ○]
└─ 5 ─┘

Think 4 + __1__ = 5.

So, 5 − 4 = __1__.

1. 3 − 2 = ?

[2 |]
└─ 3 ─┘

Think 2 + ____ = 3.

So, 3 − 2 = ____.

2. 10 − 5 = ?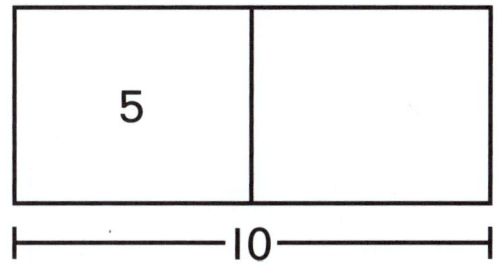

Think 5 + ____ = 10.

So, 10 − 5 = ____.

3. 9 − 5 = ?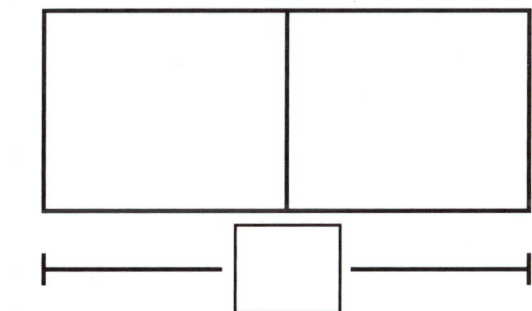

Think ____ + ____ = ____.

So, 9 − 5 = ____.

4. 7 − 6 = ?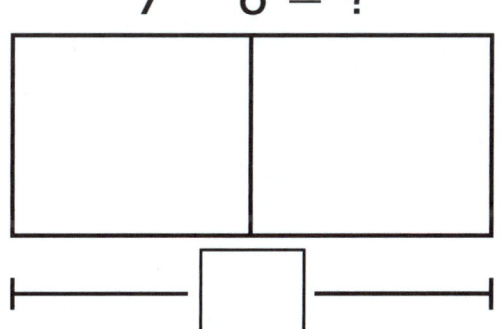

Think ____ + ____ = ____.

So, 7 − 6 = ____.

Chapter 2 | Lesson 9

one hundred seventeen 117

5. **YOU BE THE TEACHER** There are 8 birds. 5 fly away. Descartes uses addition to tell how many birds are left. Is he correct? Show how you know.

5 + 3 = 8
3 birds are left.

6. **Modeling Real Life** There are 9 kittens. 7 are adopted. How many kittens still need to be adopted?

_____ kittens

Review & Refresh

7. You have 5 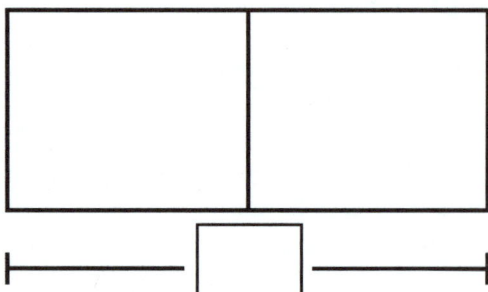 and 2 dice. How many dice do you have in all?

___ + ___ = ___

_____ dice

Name _____

Performance Task 2

1. You plant 4 red flower seeds and 4 yellow flower seeds. Your friend plants 5 red flower seeds and 4 yellow flower seeds.

   a. 1 of your yellow seeds does not grow. How many of your flowers grow?

   ____ flowers

   b. 3 of your friend's red seeds do not grow. How many of your friend's flowers grow?

   ____ flowers

   c. Who has more flowers?

         You   Friend

   d. How many more red flowers do you have than your friend?

   ____ flowers

Add or Subtract

To Play: Players take turns. On your turn, spin both spinners. Write the numbers on your Add or Subtract Recording Sheet. Decide whether you want to add or subtract the numbers you spin. Place a counter on your sum or difference. Your turn is over. Play until a player gets 4 counters in a row.

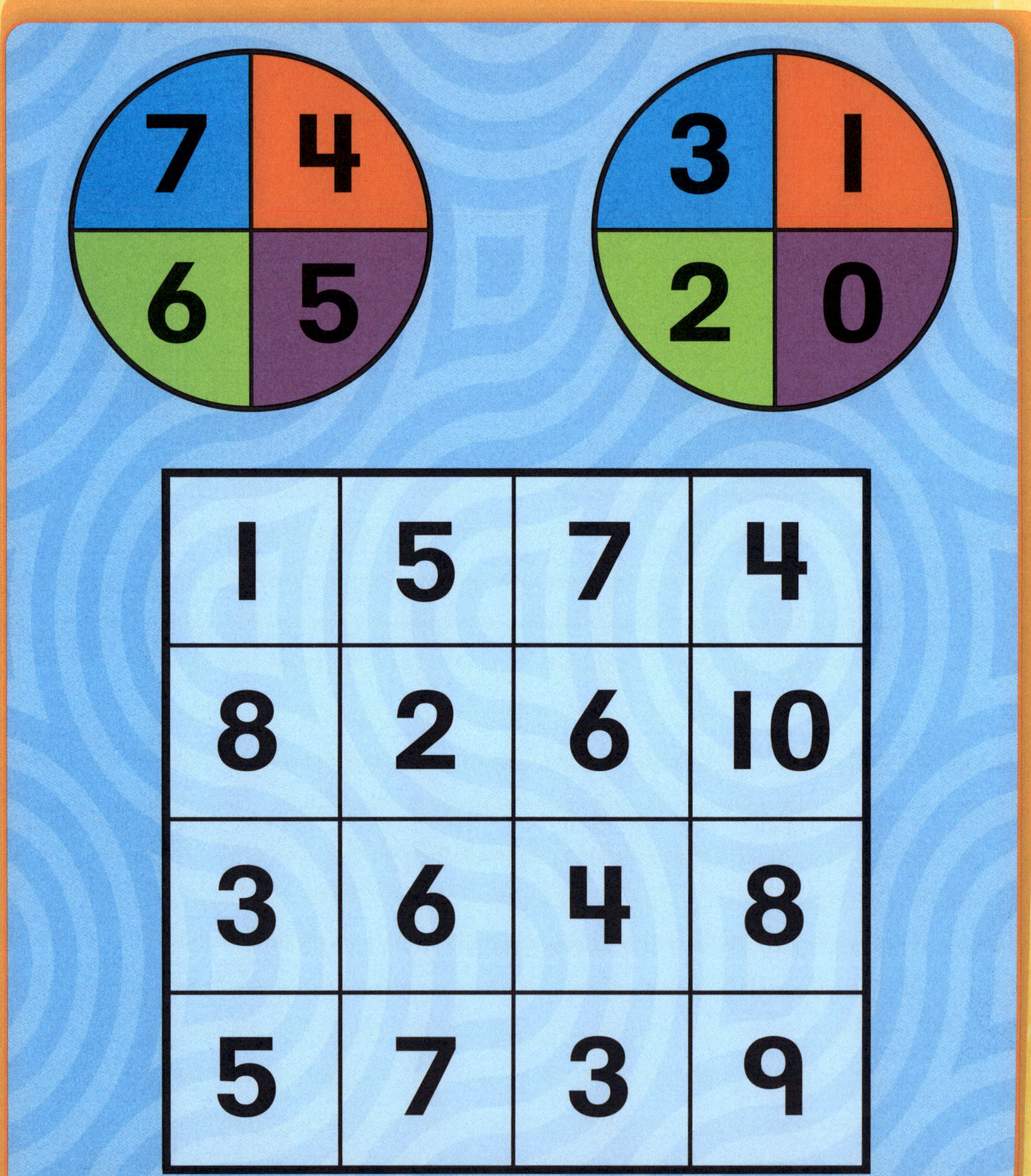

Name _____

Chapter Practice 2

2.1 Add 0

Use the picture to write an equation.

1.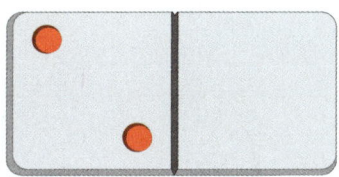

   ___ + 0 = ___

2.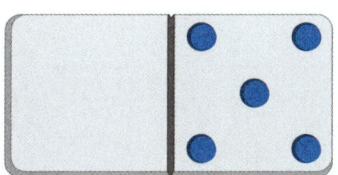

   0 + ___ = ___

3. 7 + 0 = ___

4. 0 + 6 = ___

2.2 Subtract 0 and Subtract All

Use the picture to write an equation.

5.

   ___ − 0 = ___

6.

   ___ − ___ = 0

7. 9 − 9 = ___

8. 8 − 0 = ___

2.3 Add and Subtract 1

Use the picture to write an equation.

9.

___ + ___ = ___

10.

___ − ___ = ___

11. 3 − 1 = ___

12. 5 + 1 = ___

2.4 Add Doubles from 1 to 5

13.

___ + ___ = ___

14.

15. **Reasoning** Circle the equations you can complete using doubles.

___ + ___ = 10   ___ + ___ = 3

___ + ___ = 6   ___ + ___ = 7

2.5 Use Doubles

Use the double 2 + 2 to find each sum.

16. 2 + 3 = ____

2 + 1 = ____

Find the sum. Write the double you used.

17. 4 + 5 = ____

____ + ____ = ____

18.

```
   4
+  3
____
```

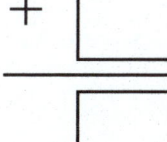

2.6 Add in Any Order

19.

____ + ____ = ____   ____ + ____ = ____

20. **Number Sense** Use the numbers shown to write two addition equations.

____ + ____ = ____

9   6   3

____ + ____ = ____

2.7 Count On to Add

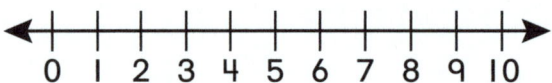

21. $5 + 2 =$ _____

22. $6 + 4 =$ _____

2.8 Count Back to Subtract

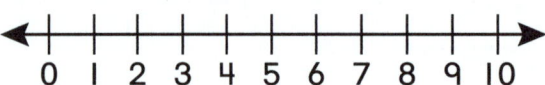

23. $9 - 7 =$ _____

24. $10 - 5 =$ _____

2.9 Use Addition to Subtract

25. $5 - 3 = ?$

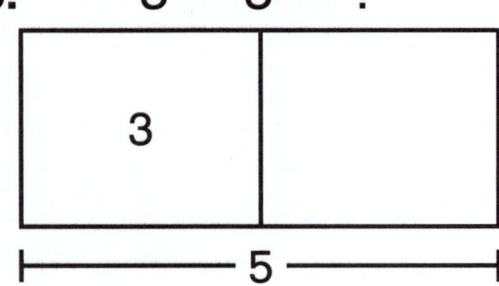

Think $3 +$ _____ $= 5$.

So, $5 - 3 =$ _____.

26. $9 - 8 = ?$

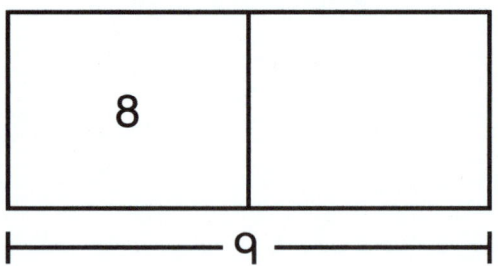

Think $8 +$ _____ $= 9$.

So, $9 - 8 =$ _____.

3 More Addition and Subtraction Situations

- Have you ever helped bake something?
- How many eggs do you see? How many more eggs do you need to have 10 eggs in all?

Chapter Learning Target:
Understand problem solving.

Chapter Success Criteria:
- I can identify problems.
- I can describe fact families.
- I can explain an equation.
- I can apply strategies.

3 Vocabulary

Review Words
count on
number line

Organize It

Use the review words to complete the graphic organizer.

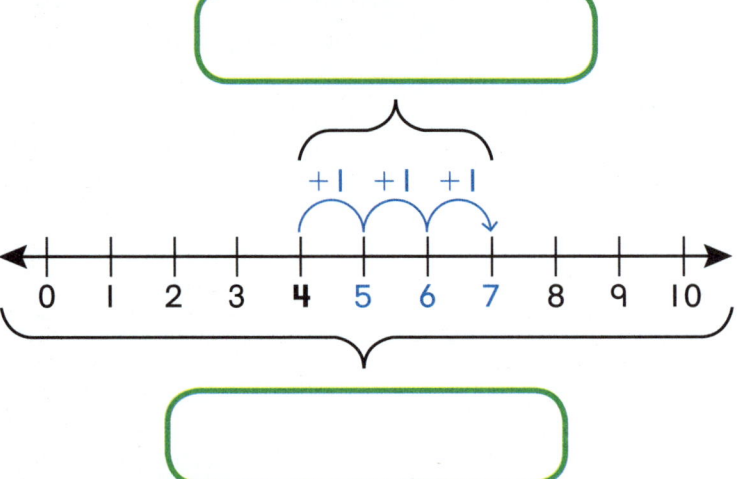

Define It

Use your vocabulary cards to identify the words.

2 + 3 = 5
3 + 2 = 5
5 − 2 = 3
5 − 3 = 2

126 one hundred twenty-six

Chapter 3 Vocabulary Cards

bar model

fact family

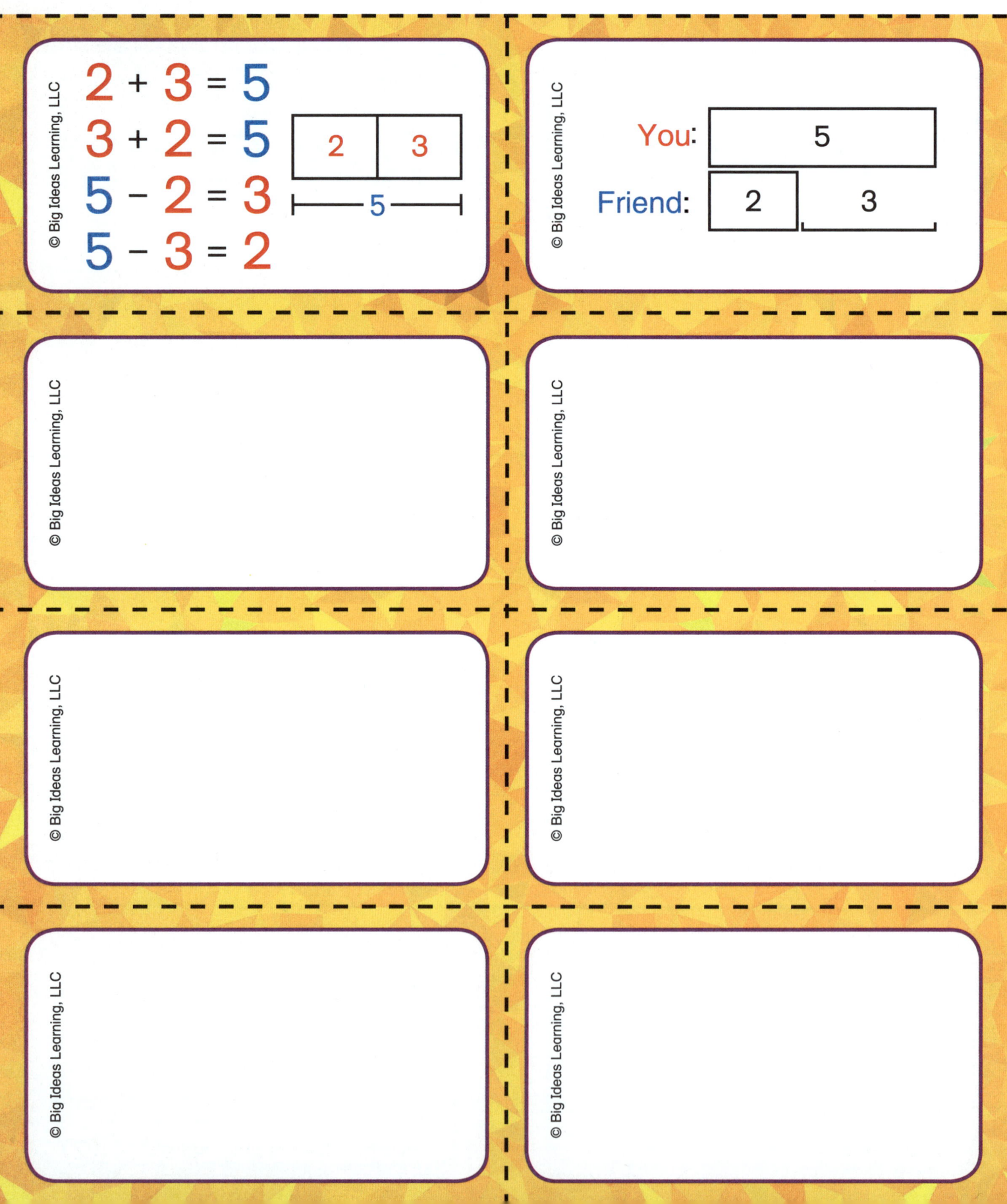

Name _____

Solve *Add To* Problems with Start Unknown   3.1

Learning Target: Solve for a missing addend given an addend and the sum.

Explore and Grow

Use counters to model each problem.

```
    2
+  ☐
───
   5
```

```
   ☐
+  2
───
   5
```

Chapter 3 | Lesson 1

one hundred twenty-seven 127

 Think and Grow

? + 4 = 6

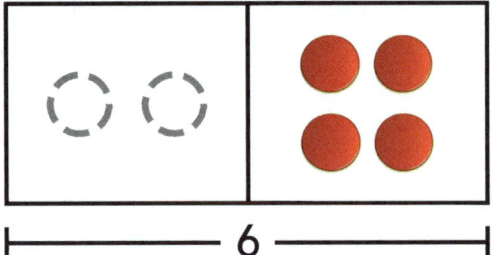

 Draw more dots to make the whole.

_2_ + 4 = 6

Show and Grow I can do it!

1. ? + 5 = 8

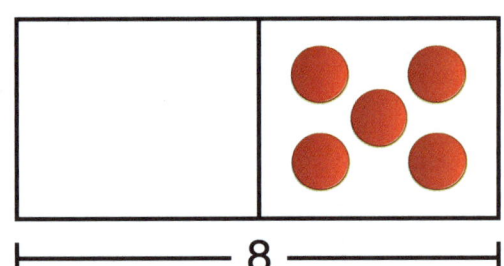

____ + 5 = 8

2. ? + 2 = 3

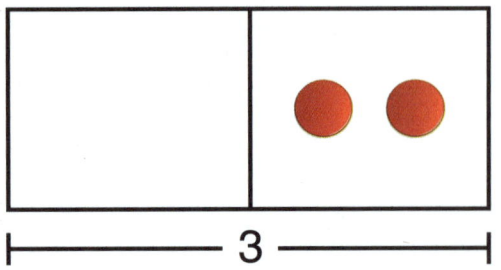

____ + 2 = 3

128 one hundred twenty-eight

Name _____

Apply and Grow: Practice

3. ? + 3 = 7

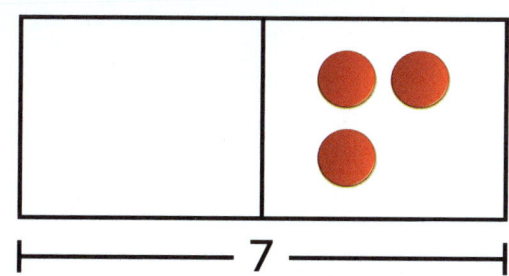

___ + 3 = 7

4. ? + 8 = 10

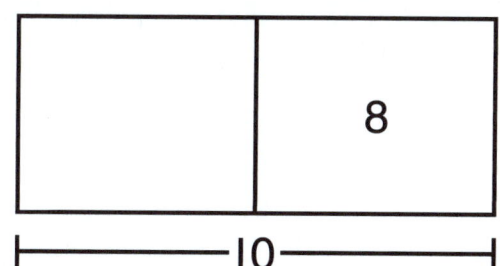

___ + 8 = 10

5. ___ + 4 = 9

6. ___ + 0 = 5

7. **Structure** Circle the equation that matches the model.

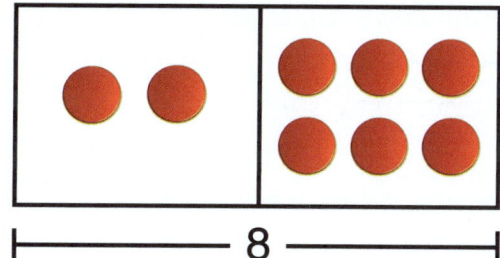

8 + 2 = 10   2 + 6 = 8

2 + 2 = 4   4 + 6 = 10

Chapter 3 | Lesson 1

one hundred twenty-nine 129

Think and Grow: Modeling Real Life

There are some ladybugs. 2 more join them. Now there are 9. How many ladybugs were there to start?

Model:

Addition equation:

_____ ladybugs

Show and Grow   I can think deeper!

8. You have some books. You get 4 more books. Now you have 10. How many books did you have to start?

Model:

Addition equation:

_____ books

Name _____

Practice 3.1

Learning Target: Solve for a missing addend given an addend and the sum.

? + 2 = 8

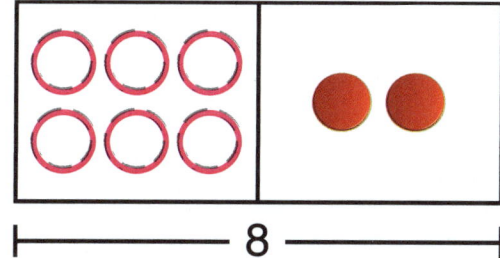

Draw more dots to make the whole.

__6__ + 2 = 8

1. ? + 5 = 10

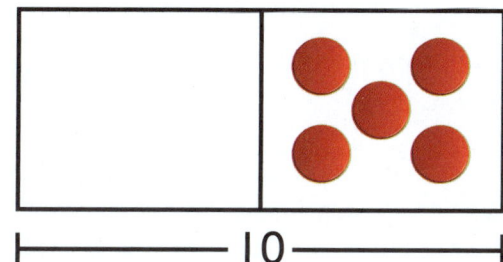

____ + 5 = 10

2. ? + 1 = 4

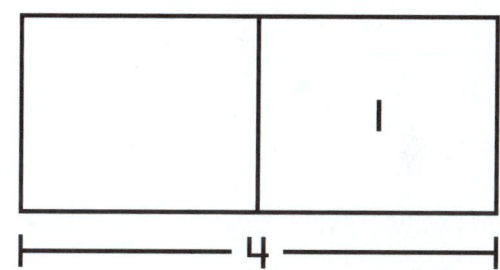

____ + 1 = 4

3. ____ + 2 = 7

4. ____ + 3 = 9

Chapter 3 | Lesson 1

one hundred thirty-one 131

5. **Structure** Circle the equation that matches the model.

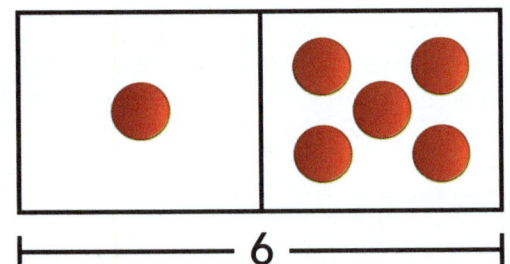

$6 + 1 = 7$ $4 + 2 = 6$

$3 + 4 = 7$ $1 + 5 = 6$

6. **Modeling Real Life** There are some hippos. 6 more join them. Now there are 9. How many hippos were there to start?

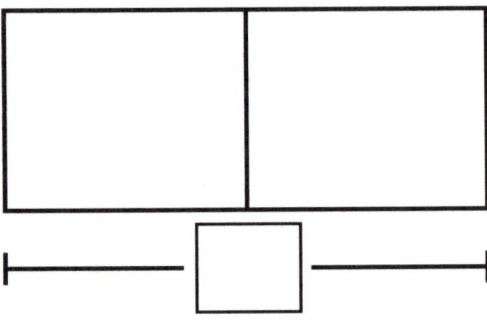

_____ hippos

Review & Refresh

7. There are 8 🍃 on a tree.
5 🍃 fall off.
How many 🍃 are left?

___ − ___ = ___

___ 🍃

8. You have 3 ⚽.
You lose 1 ⚽.
How many ⚽ do you have left?

___ − ___ = ___

___ ⚽

Name _____

Learning Target: Solve a subtraction equation to find the missing part.

Solve *Take From* Problems with Change Unknown

Explore and Grow

Use counters to model each problem.

$$\begin{array}{r} 6 \\ -\ 2 \\ \hline \Box \end{array}$$

$$\begin{array}{r} 6 \\ -\ \Box \\ \hline 2 \end{array}$$

Chapter 3 | Lesson 2   one hundred thirty-three **133**

Think and Grow

4 − ? = 1

4 − 3 = 1

Draw more dots to make the whole.

Show and Grow — I can do it!

1. 7 − ? = 2

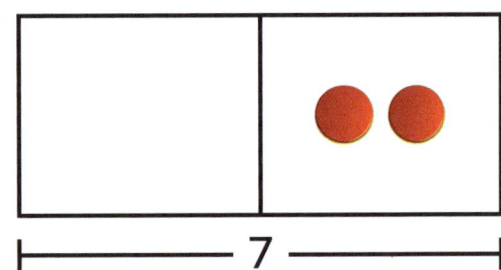

7 − ___ = 2

2. 5 − ? = 3

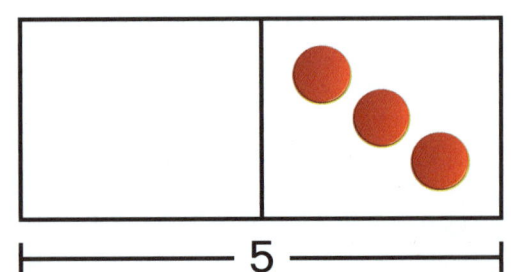

5 − ___ = 3

134 one hundred thirty-four

Apply and Grow: Practice

3. 8 − ? = 2

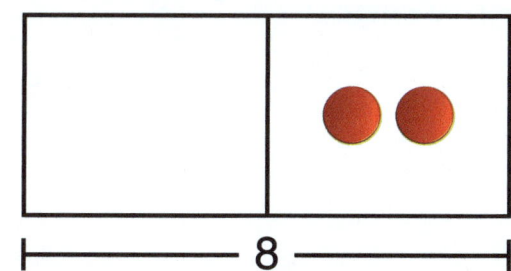

8 − ___ = 2

4. 9 − ? = 6

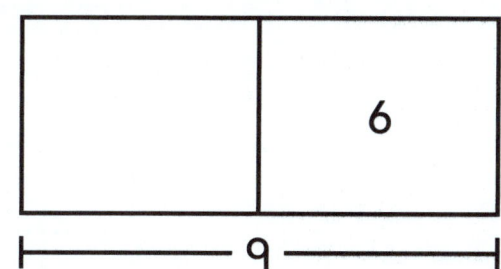

9 − ___ = 6

5. 3 − ___ = 0

6. 10 − ___ = 5

7. **Repeated Reasoning** Match each model with its correct equation.

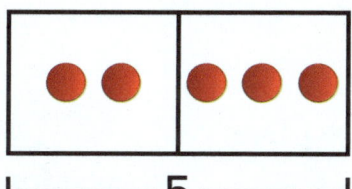

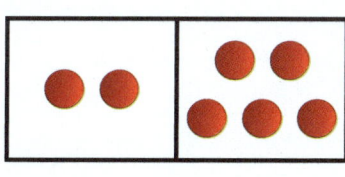

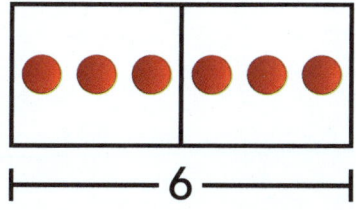

7 − 2 = 5

5 − 2 = 3

6 − 3 = 3

Chapter 3 | Lesson 2

Think and Grow: Modeling Real Life

You have 9 coins. You toss some of them into a fountain. You have 5 left. How many coins did you toss?

Model:

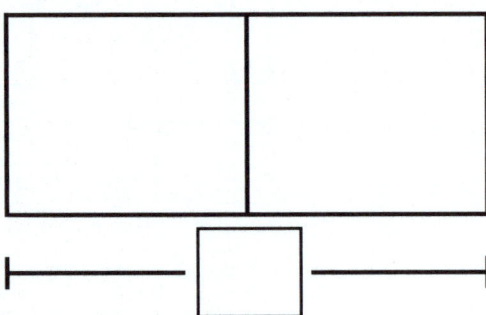

Subtraction equation: _____

_____ coins

Show and Grow — I can think deeper!

8. You have 8 crayons. You lose some of them. You have 2 left. How many crayons did you lose?

Model:

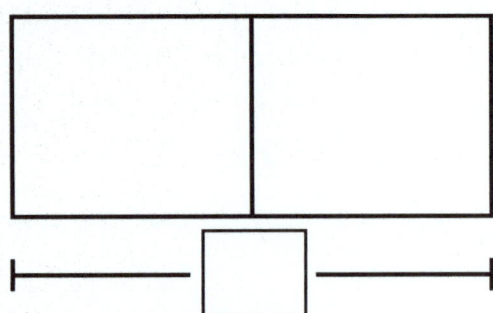

Subtraction equation: _____

_____ crayons

136 one hundred thirty-six

Name _____

Practice 3.2

Learning Target: Solve a subtraction equation to find the missing part.

$6 - ? = 2$

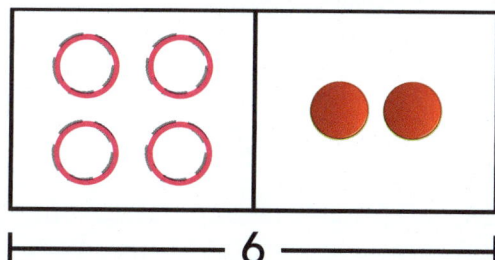

Draw more dots to make the whole.

$6 - \underline{4} = 2$

1. $8 - ? = 4$

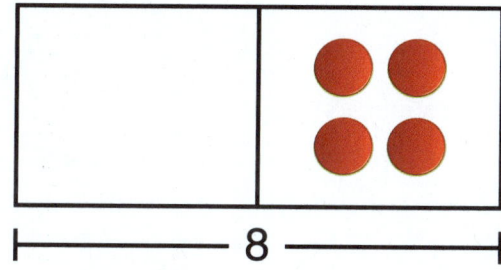

$8 - \underline{} = 4$

2. $7 - ? = 3$

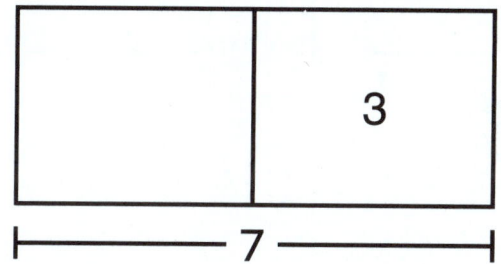

$7 - \underline{} = 3$

3. $5 - \underline{} = 4$

4. $6 - \underline{} = 6$

Chapter 3 | Lesson 2

one hundred thirty-seven 137

5. **Repeated Reasoning** Match each model with its correct equation.

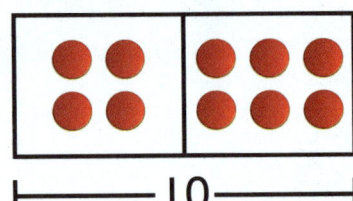

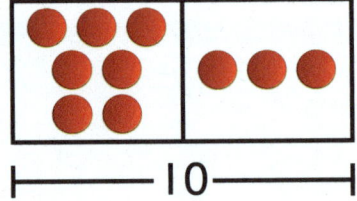

10 − 7 = 3   10 − 4 = 6

6. **Modeling Real Life** You have 10 toys. Your friend borrows some of them. You have 7 left. How many toys did your friend borrow?

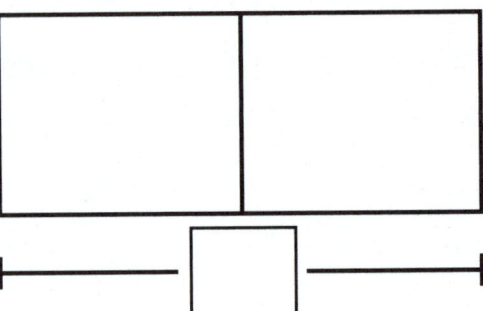

_____ toys

Review & Refresh

7. You have 6 🍌.
You buy 3 more 🍌.
How many 🍌 do you have now?

___ + ___ = ___

___ 🍌

8. You have 4 🖍.
You buy 1 more 🖍.
How many 🖍 do you have now?

___ + ___ = ___

___ 🖍

Name _____

Learning Target: Solve a subtraction equation to find the whole.

Solve Take From Problems with Start Unknown 3.3

 Explore and Grow

Use counters to model each problem.

$3 + 4 = \underline{}$

$\underline{} - 3 = 4$

Chapter 3 | Lesson 3   one hundred thirty-nine

Think and Grow

? − 3 = 2

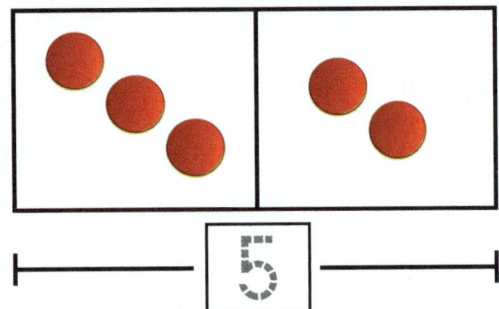

Think 3 + 2 = __5__.

So, __5__ − 3 = 2.

Add the parts to find the whole.

Show and Grow — I can do it!

1. ? − 2 = 4

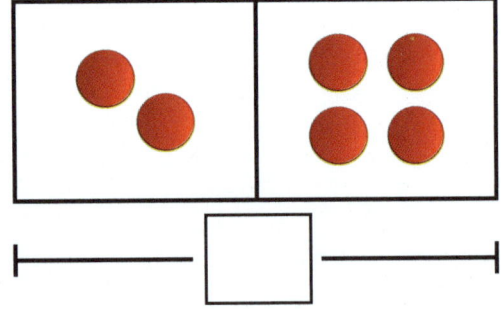

Think 2 + 4 = ____.

So, ____ − 2 = 4.

2. ? − 1 = 1

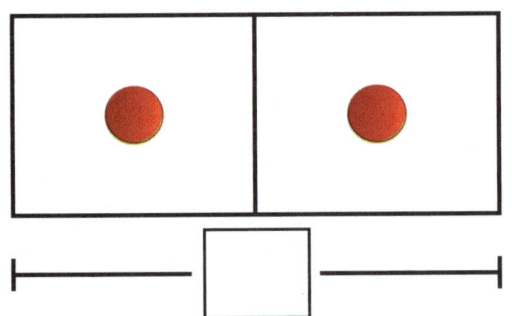

Think 1 + 1 = ____.

So, ____ − 1 = 1.

140   one hundred forty

Name _____

✓ Apply and Grow: Practice

3. ? − 2 = 5

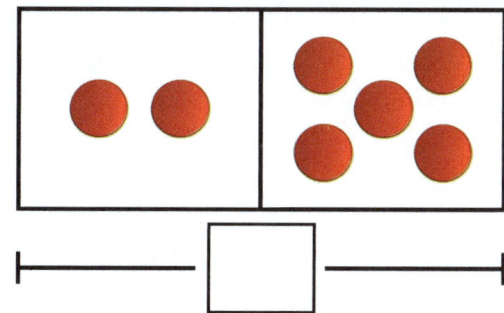

Think 2 + 5 = ___.

So, ___ − 2 = 5.

4. ? − 3 = 6

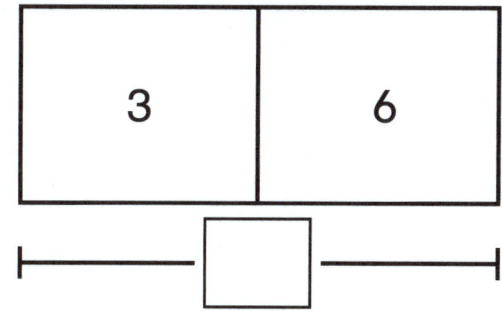

Think 3 + 6 = ___.

So, ___ − 3 = 6.

5. ? − 4 = 1

Think 4 + 1 = ___.

So, ___ − 4 = 1.

6. ? − 6 = 4

Think 6 + 4 = ___.

So, ___ − 6 = 4.

7. **MP Structure** Circle the equations that match the model.

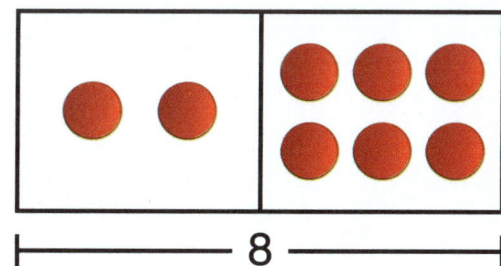

$8 - 2 = 6$ $8 + 2 = 10$

$6 - 2 = 4$ $2 + 6 = 8$

Chapter 3 | **Lesson 3**   one hundred forty-one   **141**

Think and Grow: Modeling Real Life

A group of students are at a playground. 2 of them leave. There are 8 left. How many students were there to start?

Model:

Subtraction equation:

_____ students

Show and Grow   I can think deeper!

8. You have some strawberries. You eat 9 of them. You have 0 left. How many strawberries did you have to start?

Model:

Subtraction equation:

_____ strawberries

Name _____

Practice 3.3

Learning Target: Solve a subtraction equation to find the whole.

? − 3 = 1

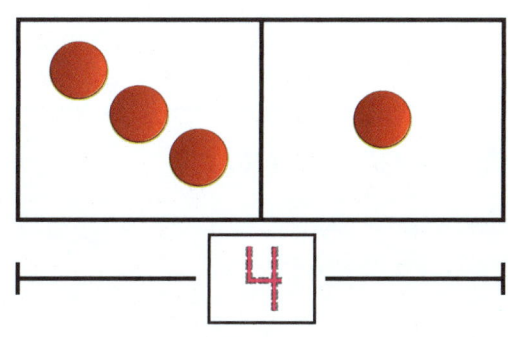

Think 3 + 1 = __4__.

So, __4__ − 3 = 1.

1. ? − 7 = 2

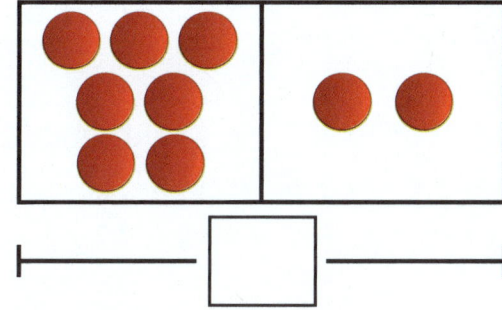

Think 7 + 2 = ____.

So, ____ − 7 = 2.

2. ? − 2 = 8

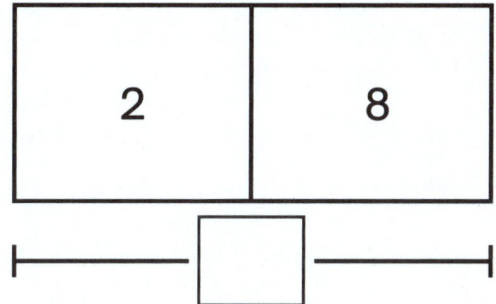

Think 2 + 8 = ____.

So, ____ − 2 = 8.

3. ? − 3 = 0

Think 3 + 0 = ____.

So, ____ − 3 = 0.

4. ? − 2 = 6

Think 2 + 6 = ____.

So, ____ − 2 = 6.

Chapter 3 | Lesson 3

one hundred forty-three 143

5. **Structure** Circle the equations that match the model.

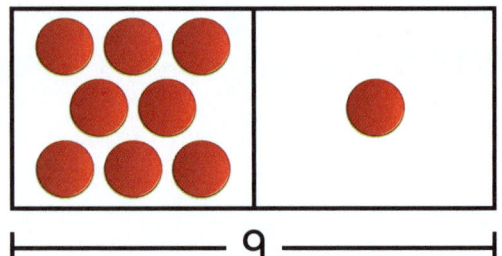

$8 - 1 = 7$ $9 + 1 = 10$

$9 - 8 = 1$ $8 + 1 = 9$

6. **Modeling Real Life** There are some people on a trolley. 4 of them exit. There are 4 people left. How many people were on the trolley to start?

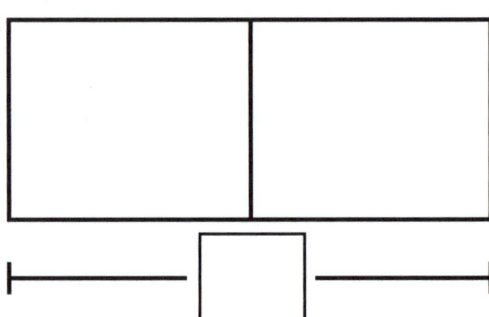

_____ people

Review & Refresh

7. There are 5 blue balloons and 3 red balloons. How many more blue balloons are there?

___ − ___ = ___

_____ more blue balloons

144   one hundred forty-four

Name _____

Learning Target: Solve *compare* word problems when given how many more.

Compare Problems: Bigger Unknown 3.4

 **Explore and Grow**

Use counters to model the story.

Newton has 5 balls. Descartes has 2 more balls than Newton. How many balls does Descartes have?

Descartes

Newton

Descartes has ____ balls.

Think and Grow

Your friend has 4 stickers. You have 1 more than your friend. How many stickers do you have?

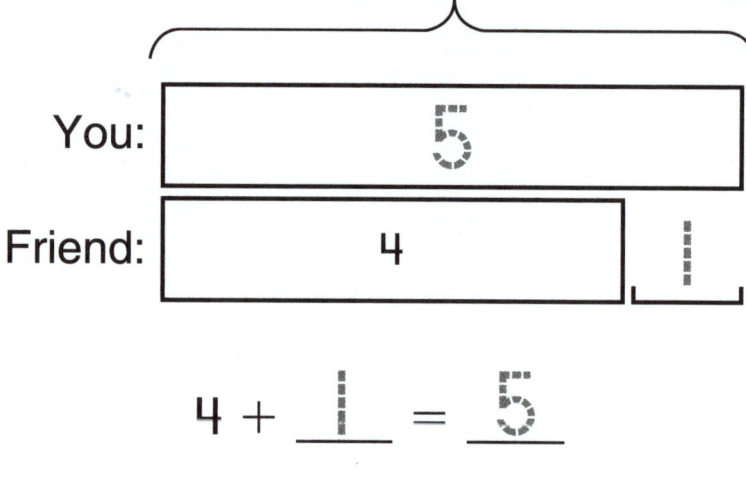

4 + __1__ = __5__

__5__ stickers

Show and Grow — I can do it!

1. Your friend has 7 trading cards. You have 3 more than your friend. How many trading cards do you have?

You:

Friend: 7

7 + ___ = ___

___ trading cards

Apply and Grow: Practice

2. Your friend has 1 soccer ball. You have 2 more than your friend. How many soccer balls do you have?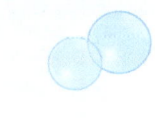

   You: [   ]
   Friend: [1]

   1 + ___ = ___

   ___ soccer balls

3. Your friend swims 4 more laps than you. You swim 3 laps. How many laps does your friend swim?

   Friend: [   ]
   You: [   ]

   ___ + ___ = ___

   ___ laps

4.  **Precision** Your friend catches 5 more fish than you. You catch 2 fish. How many fish does your friend catch? Circle the bar model that matches the problem.

   Friend: [5]   Friend: [7]
   You: [2][3]   You: [2][5]

Chapter 3 | Lesson 4

Think and Grow: Modeling Real Life

Your friend has 1 yellow flower and 2 red flowers. You have 3 more flowers than your friend. How many flowers do you have?

Model: You: [   ]

Friend: [   ][ ]

Addition equation:

_____ flowers

Show and Grow — I can think deeper!

5. Your friend has 6 gray shirts and 2 blue shirts. You have 2 more shirts than your friend. How many shirts do you have?

Model: You: [   ]

Friend: [   ][ ]

Addition equation:

_____ shirts

Name _____

Practice 3.4

Learning Target: Solve *compare* word problems when given how many more.

Your friend has 2 toy cars. You have 4 more than your friend. How many toy cars do you have?

You: | 6 |
Friend: | 2 | 4 |

$2 + \underline{4} = \underline{6}$

$\underline{6}$ toy cars

1. You have 3 key chains. Your friend has 5 more than you. How many key chains does your friend have?

Friend: | |
You: | 3 | |

$3 + \underline{} = \underline{}$

$\underline{}$ key chains

2. You have 8 more bracelets than your friend. Your friend has 2 bracelets. How many bracelets do you have?

You: | |
Friend: | | |

$\underline{} + \underline{} = \underline{}$

$\underline{}$ bracelets

Chapter 3 | Lesson 4   one hundred forty-nine   149

3. **Precision** You have 1 seashell. Your friend has 8 more than you. How many seashells does your friend have? Circle the bar model that matches the problem.

Friend:	9		Friend:	8
You: 1	8		You: 1	7

4. **Modeling Real Life** Your friend has 2 comic books and 2 mystery books. You have 3 more books than your friend. How many books do you have?

You:
Friend:

_____ books

Review & Refresh

5. There are 2 blue crayons and 6 red crayons. How many fewer blue crayons are there?

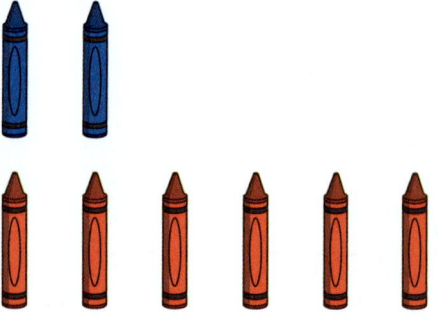

_____ − _____ = _____

_____ fewer blue crayons

Name _____

Learning Target: Solve *compare* word problems when given how many fewer.

Compare Problems: Smaller Unknown 3.5

Explore and Grow

Use counters to model the story.

Newton has 5 treats. Descartes has 2 fewer treats than Newton. How many treats does Descartes have?

Newton

Descartes

Descartes has ____ treats.

Chapter 3 | Lesson 5

 Think and Grow

Your friend builds 6 sand castles. You build 4 fewer than your friend. How many sand castles do you build?

Friend: | 6 |

You: | 2 | 4 |

 You can subtract *or* add to find the missing part!

6 − 4 = 2

2 + 4 = 6

2 sand castles

Show and Grow — I can do it!

1. Your friend has 8 stones. You have 1 fewer than your friend. How many stones do you have?

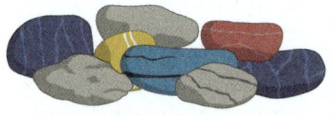

Friend: | |

You: | | 1 |

___ − ___ = ___

___ + ___ = ___

___ stones

Name _____

✓ Apply and Grow: Practice

2. You blow 5 bubbles. Your friend blows 2 fewer than you. How many bubbles does your friend blow?

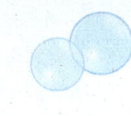

You: [   ]

Friend: [| 2]

___ + ___ = ___

___ − ___ = ___

___ bubbles

3. You have 3 fewer oranges than your friend. Your friend has 9 oranges. How many oranges do you have?

Friend: [   ]

You: [   ]

___ ◯ ___ = ___

___ oranges

4.  Complete the bar model. Do both equations match the bar model?

You: [8]

Friend: [| 3]

$8 - 3 = 5$

$5 + 3 = 8$

Yes   No

Chapter 3 | Lesson 5

Think and Grow: Modeling Real Life

Your friend has 2 blue markers and 7 yellow markers. You have 5 fewer markers than your friend. How many markers do you have?

Model: Friend:

You:

Equation:

_____ markers

Show and Grow   I can think deeper!

5. Your friend has 6 tennis balls and 1 baseball. You have 2 fewer balls than your friend. How many balls do you have?

Model: Friend:

You:

Equation:

_____ balls

Name _____

Practice 3.5

Learning Target: Solve *compare* word problems when given how many fewer.

You have 7 pencils. Your friend has 6 fewer than you. How many pencils does your friend have?

You: | 7 |
Friend: | 1 | 6 |

7 − 6 = 1

1 + 6 = 7

__1__ pencil

1. Your friend has 9 awards. You have 5 fewer than your friend. How many awards do you have?

Friend:
You: | | 5 |

___ − ___ = ___

___ + ___ = ___

____ awards

2. Your friend finds 2 fewer bugs than you. You find 4 bugs. How many bugs does your friend find?

You:
Friend:

___ ◯ ___ = ___

____ bugs

Chapter 3 | Lesson 5

3. **Reasoning** Complete the bar model. Circle the equation that matches the bar model.

You: | 7 |
Friend: | 4 |

7 − 4 = 3

7 + 3 = 10

4. **Modeling Real Life** Your friend has 8 black cats and 2 orange cats. You have 7 fewer cats than your friend. How many cats do you have?

Friend: | |
You: | |

_____ cats

Review & Refresh

5. Write the numbers of shirts and shorts. Are the numbers equal? Circle the thumbs up for *yes* or the thumbs down for *no*.

Name _____

Learning Target: Identify whether an equation is true or false.

True or False Equations 3.6

Explore and Grow

Color the flowers that have a sum or difference of 6.

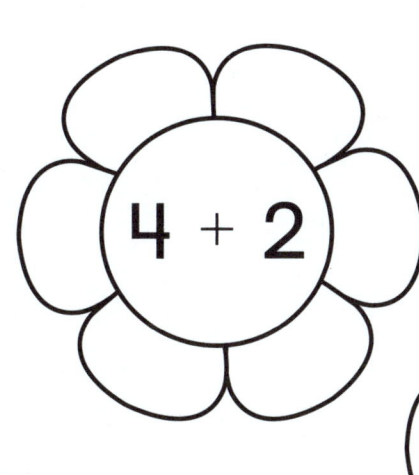

Chapter 3 | **Lesson 6**

one hundred fifty-seven 157

 Think and Grow

$7 + 1 \stackrel{?}{=} 4 + 4$

7 + 1: ◯◯◯◯◯◯◯◯

4 + 4: ◯◯◯◯◯◯◯◯

Both sides equal 8.

$8 \stackrel{?}{=} 8$

(True)   False

8 = 8 is true.

Show and Grow — I can do it!

Is the equation true or false?

1. $2 + 0 \stackrel{?}{=} 1 + 2$

   2 + 0:

   1 + 2:

   ___ $\stackrel{?}{=}$ ___

   True   False

2. $5 - 4 \stackrel{?}{=} 6 - 5$

   5 − 4:

   6 − 5:

   ___ $\stackrel{?}{=}$ ___

   True   False

Name _____

 **Apply and Grow: Practice**

Is the equation true or false?

3. $4 + 1 \stackrel{?}{=} 2 + 2$

   $4 + 1:$

   $2 + 2:$

   $\underline{} \stackrel{?}{=} \underline{}$

   True   False

4. $7 + 2 \stackrel{?}{=} 3 + 6$

   $7 + 2:$

   $3 + 6:$

   $\underline{} \stackrel{?}{=} \underline{}$

   True   False

5. $10 - 4 \stackrel{?}{=} 6 - 0$

   $\underline{} \stackrel{?}{=} \underline{}$

   True   False

6. $5 - 2 \stackrel{?}{=} 7 - 4$

   $\underline{} \stackrel{?}{=} \underline{}$

   True   False

7. $6 + 1 \stackrel{?}{=} 6 - 1$

   $\underline{} \stackrel{?}{=} \underline{}$

   True   False

8. $3 - 1 \stackrel{?}{=} 1 + 2$

   $\underline{} \stackrel{?}{=} \underline{}$

   True   False

9. **Number Sense** Circle all of the equations that are true.

   $7 \stackrel{?}{=} 7$ $5 + 5 \stackrel{?}{=} 6 + 4$

   $10 - 4 \stackrel{?}{=} 8$ $4 + 2 \stackrel{?}{=} 9 - 3$

Chapter 3 | Lesson 6

Think and Grow: Modeling Real Life

You have 7 marbles. You lose 2 of them. Your friend has 4 marbles and finds 3 more. Do you and your friend have the same number of marbles?

Equation: ___ − ___ $\overset{?}{=}$ ___ + ___

___ $\overset{?}{=}$ ___

Yes   No

Show and Grow   I can think deeper!

10. You have 1 balloon. You blow up 3 more. Your friend has 5 balloons. 1 of your friend's balloons pops. Do you and your friend have the same number of balloons?

Equation: ___ + ___ $\overset{?}{=}$ ___ − ___

___ $\overset{?}{=}$ ___

Yes   No

Name _____

Practice 3.6

Learning Target: Identify whether an equation is true or false.

$5 + 2 \stackrel{?}{=} 3 + 3$

5 + 2: ◯ ◯ ◯ ◯ ◯ ◯ ◯

3 + 3: ◯ ◯ ◯ ◯ ◯ ◯

__7__ $\stackrel{?}{=}$ __6__   True   (False)

Is the equation true or false?

1. $7 + 0 \stackrel{?}{=} 2 + 4$

   7 + 0:

   2 + 4:

   ___ $\stackrel{?}{=}$ ___

   True   False

2. $8 - 1 \stackrel{?}{=} 9 - 2$

   8 − 1:

   9 − 2:

   ___ $\stackrel{?}{=}$ ___

   True   False

3. $9 - 5 \stackrel{?}{=} 1 + 3$

   ___ $\stackrel{?}{=}$ ___

   True   False

4. $8 - 1 \stackrel{?}{=} 4 + 4$

   ___ $\stackrel{?}{=}$ ___

   True   False

Chapter 3 | Lesson 6   one hundred sixty-one   161

5. **Number Sense** Circle all of the equations that are false.

$2 \stackrel{?}{=} 4$ $4 + 5 \stackrel{?}{=} 6 - 2$

$4 + 4 \stackrel{?}{=} 10 - 2$ $9 \stackrel{?}{=} 9 - 0$

6. **Modeling Real Life** You have 5 crayons. You find 3 more. Your friend has 7 crayons and finds 1 more. Do you and your friend have the same number of crayons?

___ + ___ $\stackrel{?}{=}$ ___ + ___

___ $\stackrel{?}{=}$ ___

Yes   No

Review & Refresh

7. Circle the triangles.

   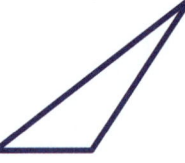

 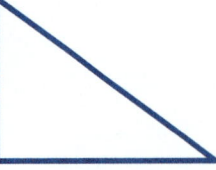  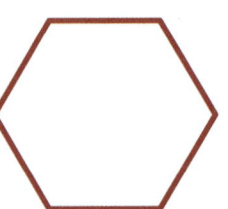

Name _____

Learning Target: Find the missing addend that makes 10.

Find Numbers That Make 10 — 3.7

Explore and Grow

Place some red counters on the ten frame. Add yellow counters to fill the frame. Write an equation to match.

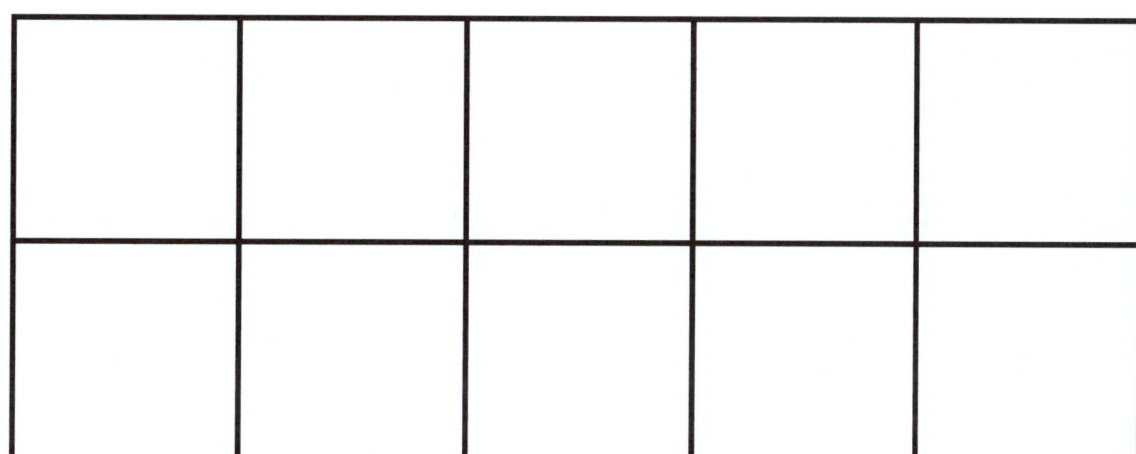

___ + ___ = ___

Chapter 3 | Lesson 7

Think and Grow

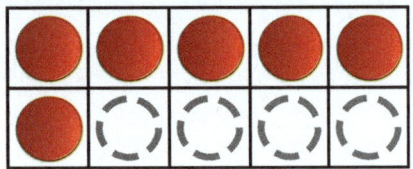

Draw 4 more circles to make 10.

6 + __4__ = 10

Show and Grow — I can do it!

1.

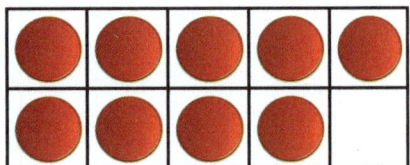

9 + ___ = 10

2.

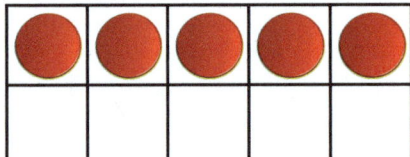

5 + ___ = 10

3.

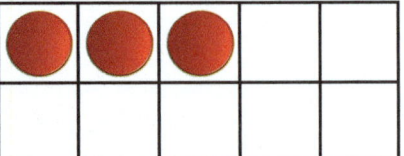

3 + ___ = 10

4.

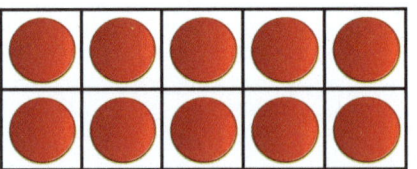

10 + ___ = 10

Name _____

Apply and Grow: Practice

5.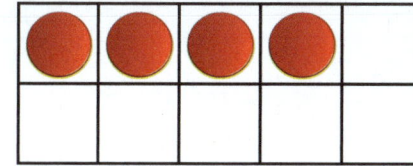

$4 + \_\_\_ = 10$

6.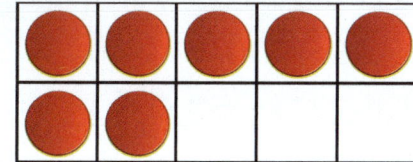

$7 + \_\_\_ = 10$

7. $1 + \_\_\_ = 10$

8. $8 + \_\_\_ = 10$

9. $\_\_\_ + 2 = 10$

10. $\_\_\_ + 5 = 10$

11. $\_\_\_ + 3 = 10$

12. $\_\_\_ + 0 = 10$

13. **DIG DEEPER!** Match the numbers that have a sum of 10.

4   3   5   1

5   7   6   9

Chapter 3 | Lesson 7

Think and Grow: Modeling Real Life

There are 7 jump ropes. Your teacher buys some more. Now there are 10. How many jump ropes did your teacher buy?

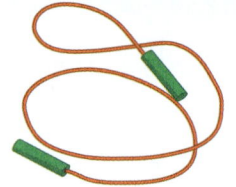

Model:

Addition equation:

_____ jump ropes

Show and Grow   I can think deeper!

14. There are 2 penguins. Some more join them. Now there are 10. How many more penguins joined them?

    Model:

    Addition equation:

    _____ penguins

Name _____

Practice

Learning Target: Find the missing addend that makes 10.

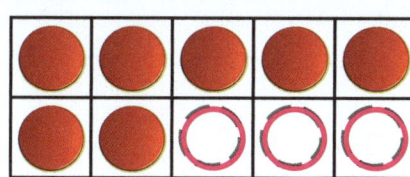

Draw 3 more circles to make 10.

7 + __3__ = 10

1.

   8 + ____ = 10

2.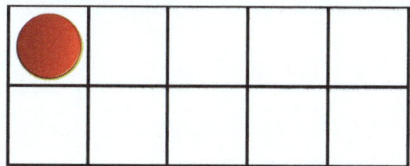

   1 + ____ = 10

3. 10 + ____ = 10

4. 5 + ____ = 10

5. ____ + 4 = 10

6. ____ + 0 = 10

Chapter 3 | Lesson 7   one hundred sixty-seven   167

7. **DIG DEEPER!** Match the numbers that have a sum of 10.

   2   8
   4   9
   3   7
   1   6

8. **Modeling Real Life** You have 3 baseball cards. Your friend gives you some more. Now you have 10. How many baseball cards did your friend give you?

   _____ baseball cards

   Review & Refresh

   Find the sum. Then change the order of the addends. Write the new addition problem.

   9.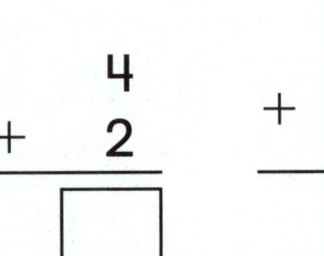

   10.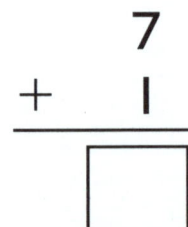

Name _____

Fact Families 3.8

Learning Target: Write related addition and subtraction equations to complete a fact family.

 **Explore and Grow**

Use linking cubes to model the equations.

4 + 5 = ___   9 − 4 = ___

5 + 4 = ___   9 − 5 = ___

Chapter 3 | Lesson 8

Think and Grow

Add the parts in any order.

Subract each part from the whole.

| 2 | 3 |

⊢——— 5 ———⊣

fact family

2 + 3 = 5   5 − 2 = 3

3 + 2 = 5   5 − 3 = 2

Show and Grow — I can do it!

1. Complete the fact family.

| 1 | 8 |

⊢——— 9 ———⊣

1 + 8 = ___   9 − ___ = ___

___ + ___ = ___   9 − ___ = ___

170 one hundred seventy

Apply and Grow: Practice

Complete the fact family.

2.

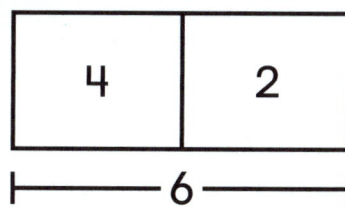

4 + 2 = ___   6 – ___ = ___

___ + ___ = ___   6 – ___ = ___

3.

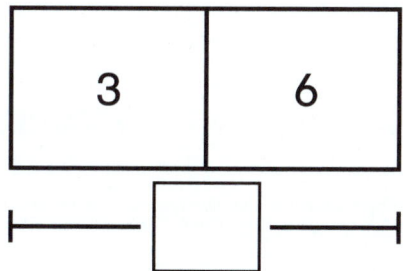

3 + 6 = ___   ___ – 6 = 3

___ + ___ = ___   ___ – ___ = ___

4. 7 + 1 = ___   ___ – 7 = ___

___ + ___ = ___   ___ – 1 = ___

5. **DIG DEEPER!** Cross out the equation that does *not* belong in the fact family.

5 + 3 = 8   5 – 3 = 2

3 + 5 = 8   8 – 5 = 3

Chapter 3 | Lesson 8

Think and Grow: Modeling Real Life

You have 3 puppets. Your friend has 7 puppets. How many fewer puppets do you have?

Model:

Equation:

_____ fewer puppets

Show and Grow — I can think deeper!

6. There are 2 spoons and 8 forks. How many more forks are there?

Model:

Equation:

_____ more forks

Name _____

Practice 3.8

Learning Target: Write related addition and subtraction equations to complete a fact family.

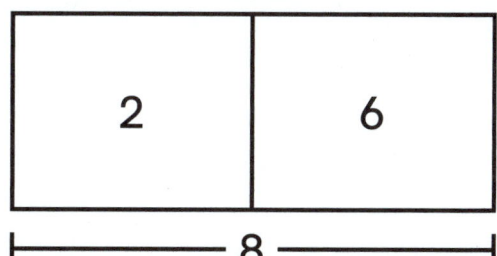

2 + 6 = _8_   8 − _6_ = _2_

_6_ + _2_ = _8_   8 − _2_ = _6_

Complete the fact family.

1.

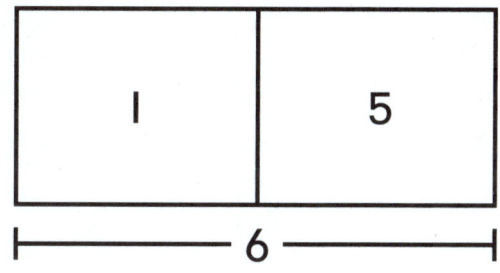

1 + 5 = ___   6 − ___ = ___

___ + ___ = ___   6 − ___ = ___

2. 3 + 7 = ___   ___ − 7 = 3

___ + ___ = ___   ___ − ___ = ___

Chapter 3 | Lesson 8 one hundred seventy-three

3. Complete the fact family.

3 + 0 = ___   ___ − 3 = ___

___ + ___ = ___   ___ − 0 = ___

4. **DIG DEEPER!** Cross out the equation that does *not* belong in the fact family.

6 + 4 = 10   10 − 6 = 4

4 + 6 = 10   6 − 4 = 2

5. **Modeling Real Life** There are 7 fish and 2 frogs. How many fewer frogs are there?

___ fewer frogs

Review & Refresh

Circle the object that holds more.

6.     7.

174   one hundred seventy-four

Name _____

Performance Task 3

1. You and your friend bake banana bread and raisin bread. You have 8 loaves of bread. 3 of them are raisin bread. Your friend has 10 loaves of bread. 6 of them are banana bread. How many more loaves of banana bread does your friend have than you?

_____ loaf

2. You give away 3 loaves of banana bread and 3 loaves of raisin bread. Your friend gives away 1 more loaf of bread than you. How many loaves of bread does your friend give away?

_____ loaves

3. You and your friend make boxes of muffins. Does each box have the same number of muffins?

Yes   No

Number Land

To Play: Put the Addition and Subtraction Cards in a pile. Start at Newton. Take turns drawing a card and moving your piece to the missing number in the equation. Repeat this process until a player gets back to Newton.

Name _____

Chapter 3 Practice

3.1 Solve *Add To* Problems with Start Unknown

1. ? + 4 = 6

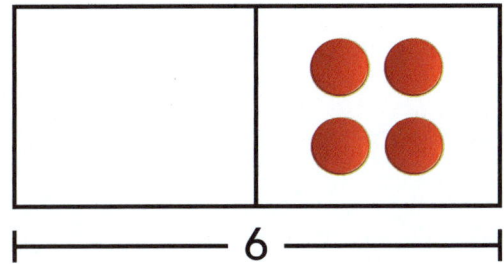

___ + 4 = 6

2. ? + 2 = 8

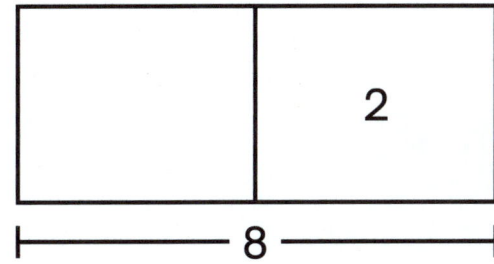

___ + 2 = 8

3.2 Solve *Take From* Problems with Change Unknown

3. 5 − ? = 4

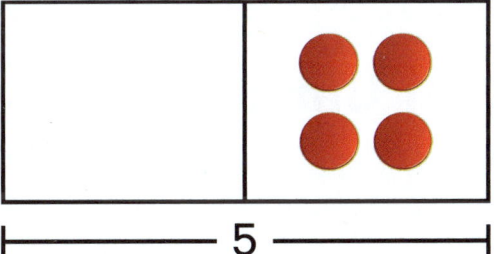

5 − ___ = 4

4. 7 − ? = 7

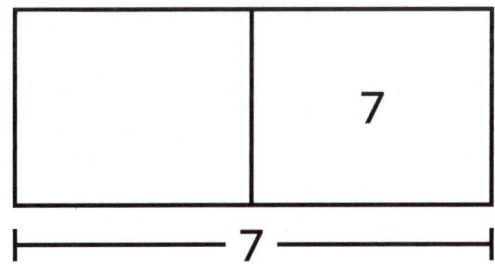

7 − ___ = 7

3.3 Solve *Take From* Problems with Start Unknown

5. ? − 6 = 3

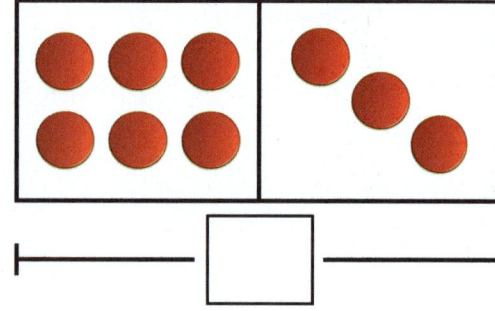

Think 6 + 3 = ____.

So, ____ − 6 = 3.

6. ? − 3 = 1

Think 3 + 1 = ____.

So, ____ − 3 = 1.

7. **MP Structure** Circle the equation that matches the model.

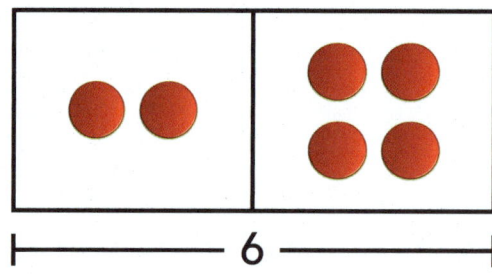

6 − 6 = 0   4 − 2 = 2   6 − 2 = 4

3.4 *Compare* Problems: Bigger Unknown

8. Your friend has 3 stickers. You have 4 more than your friend. How many stickers do you have?

You: [   ]

Friend: [3]

3 + ____ = ____

____ stickers

178   one hundred seventy-eight

3.5 Compare Problems: Smaller Unknown

9. Your friend has 5 stuffed animals. You have 2 fewer than your friend. How many stuffed animals do you have?

Friend:

You:

___ ◯ ___ = ___

___ stuffed animals

10. Modeling Real Life Your friend has 4 dogs and 2 cats. You have 1 fewer pet than your friend. How many pets do you have?

Friend:

You:

___ pets

3.6 True or False Equations

Is the equation true or false?

11. $5 + 3 \stackrel{?}{=} 10 - 2$

___ $\stackrel{?}{=}$ ___

True   False

12. $7 - 1 \stackrel{?}{=} 6 + 1$

___ $\stackrel{?}{=}$ ___

True   False

13.  **Number Sense** Circle all of the equations that are true.

$$8 \overset{?}{=} 2 \qquad\qquad 1 + 1 \overset{?}{=} 4 - 2$$

$$6 + 4 \overset{?}{=} 10 \qquad\qquad 3 + 2 \overset{?}{=} 3 + 3$$

3.7 Find Numbers That Make 10

14.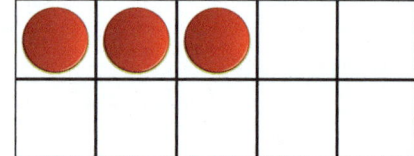

   $3 + \underline{} = 10$

15.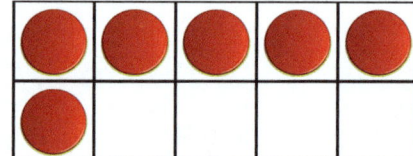

   $6 + \underline{} = 10$

3.8 Fact Families

16. Complete the fact family.

   $8 + 1 = \underline{} \qquad\qquad \underline{} - 8 = 1$

   $\underline{} + \underline{} = \underline{} \qquad\qquad \underline{} - \underline{} = \underline{}$

17. **Modeling Real Life** There are 2 slides and 6 swings on a playground. How many more swings are there?

   ____ more swings

Name _____

Cumulative Practice 1-3

1. Shade the circle next to the answer.

   4 + 4 = ___

   ○ 4 ○ 6
   ○ 9 ○ 8

2. Shade the circle next to the addition equation that you can use to solve 8 − 3.

   ○ 8 + 3 = 11

   ○ 3 + 5 = 8

   ○ 1 + 8 = 9

   ○ 5 + 2 = 7

3. Circle the equation that matches the bar model.

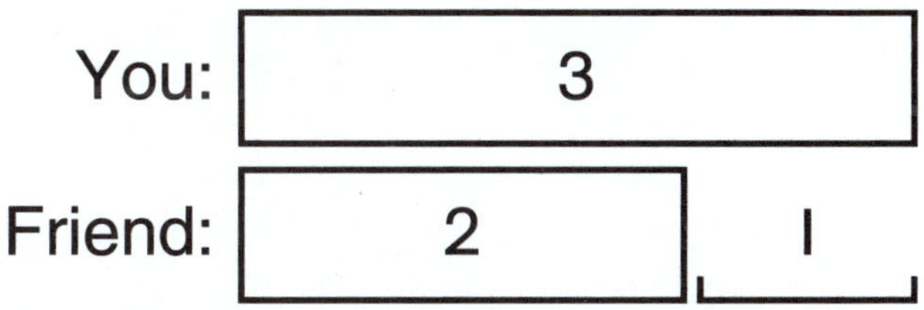

   You: | 3 |
   Friend: | 2 | 1 |

   3 + 1 = 4

   2 + 1 = 3

Chapter 3

4. You take 10 pictures. Your friend takes 3 pictures. Shade the circle next to the equation that shows how many more pictures you take.

○ 3 + 3 = 6

○ 10 − 3 = 7

○ 10 + 3 = 13

○ 3 − 1 = 2

5. Shade the circle next to the number that completes the addition equation.

___ + 7 = 9

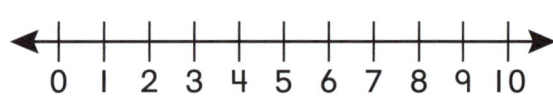

○ 1

○ 2

○ 3

○ 4

6. There are 6 🐠.

3 more 🐠 join them.

How many 🐠 are there now?

___ + ___ = ___

___ 🐠

182   one hundred eight-two

7. Is each equation true or false?

$0 + 1 \stackrel{?}{=} 0 + 8$	True	False
$1 + 5 \stackrel{?}{=} 9 - 5$	True	False
$10 - 8 \stackrel{?}{=} 7 - 5$	True	False
$4 - 3 \stackrel{?}{=} 4 + 3$	True	False

8. Use the picture to write a subtraction equation.

____ − 0 = ____

9. You have 8 beads. 5 are orange. The rest are blue. Shade the circles next to the equations that describe the beads.

○ $3 + 5 = 8$

○ $8 - 2 = 6$

○ $4 + 4 = 8$

○ $8 - 5 = 3$

Chapter 3

10. Use the numbers shown to write two equations.

             3   8   5

___ + ___ = ___   ___ + ___ = ___

11. Shade the circle next to the equation that does *not* belong in the fact family.

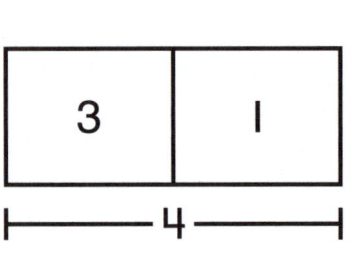

○ 3 + 1 = 4

○ 4 − 3 = 1

○ 1 + 3 = 4

○ 3 − 1 = 2

12. There are 3 rabbits. 3 more join them. Shade the circle next to the equation that shows how many rabbits there are in all.

○ 3 + 2 = 5

○ 3 + 4 = 7

○ 4 + 1 = 5

○ 3 + 3 = 6

184   one hundred eighty-four

Add Numbers within 20

Chapter Learning Target:
Understand counting strategies.

Chapter Success Criteria:
- I can identify counting strategies.
- I can describe equations.
- I can explain the strategy I used.
- I can apply strategies to solve word problems.

- How would you describe the weather today?
- How many sunny or cloudy days do you see in the forecast?

one hundred eighty-five 185

Name _____

 Vocabulary

Review Words
addend
doubles
doubles minus 1
doubles plus 1
sum

Organize It

Use the review words to complete the graphic organizer.

Define It

Match the review word to its definition.

1. addend $5 + 3 = 8$

2. sum $4 + 3 = 7$

186 one hundred eighty-six

Name _____

Learning Target: Find the sum of doubles from 6 to 10.

Use counters to model the story.

You have 7 marbles. Your friend has 7 marbles. How many marbles are there in all?

____ marbles

Chapter 4 | Lesson 1

 **Think and Grow**

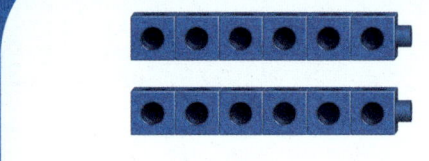

$\underline{6} + \underline{6} = \underline{12}$

 The addends are the same, so these are doubles.

$\begin{array}{r} \boxed{7} \\ + \boxed{7} \\ \hline \boxed{14} \end{array}$

Show and Grow — I can do it!

1. ___ + ___ = ___

2. ___ + ___ = ___

3.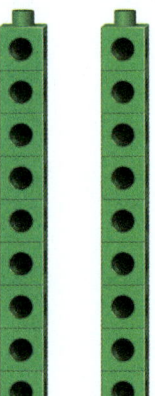

4.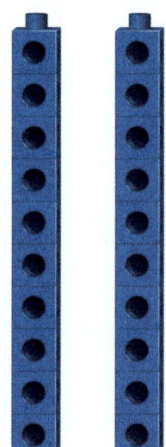

Name _____

✓ Apply and Grow: Practice

5.

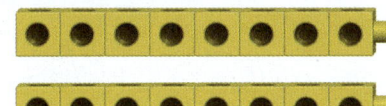

___ + ___ = ___

6.

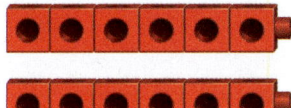

___ + ___ = ___

7. 7 + 7 = ___

8. 10 + 10 = ___

9.   9
  + 9
  ☐

10.   8
  + 8
  ☐

11. **Reasoning** You and your friend each read 6 books. How many books do you and your friend read in all?

____ books

 **Think and Grow: Modeling Real Life**

You and your friend have the same number of video games. There are 16 in all. How many video games do you have?

Draw a picture:

Addition equation:

_____ video games

Show and Grow — I can think deeper!

12. 2 friends give you the same number of pictures. You have 12 in all. How many pictures does each friend give you?

Draw a picture:

Addition equation:

_____ pictures

Name _____

Practice 4.1

Learning Target: Find the sum of doubles from 6 to 10.

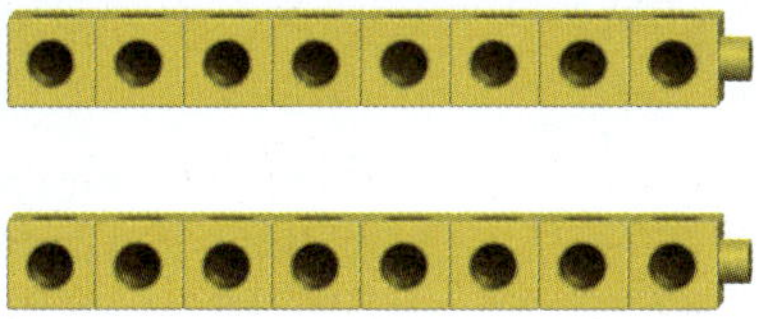

8 + 8 = 16

7 + 7 = 14

1. ___ + ___ = ___

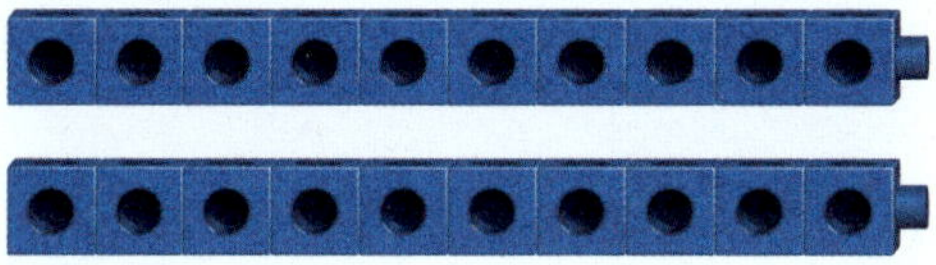

2. ___ + ___ = ___

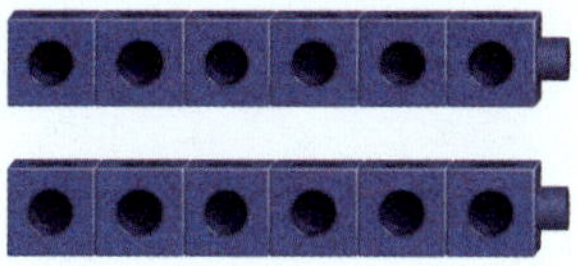

3. 9
  + 9
  ☐

4. 10
  + 10
  ☐

Chapter 4 | Lesson 1   one hundred ninety-one   191

5. **Reasoning** You and your friend each do 7 jumping jacks. How many jumping jacks do you and your friend do in all?

_____ jumping jacks

6. **Modeling Real Life** Newton and Descartes each have the same number of treats. They have 18 treats in all. How many treats does Newton have?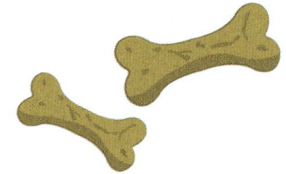

_____ treats

Review & Refresh

Use the picture to write an equation.

7.

___ + ___ = ___

8.

___ + ___ = ___

Name _____

Learning Target: Use the *doubles plus 1* and *doubles minus 1* strategies to find a sum.

Use counters to model the story.

You collect 8 leaves. Your friend collects 8 leaves. How many leaves are there in all?

_____ leaves

You collect 1 more leaf. How many leaves are there now?

_____ leaves

Chapter 4 | Lesson 2

Think and Grow

Use the double 8 + 8 to find each sum.

8 + 9 = __17__

doubles plus 1

8 + 9 is equal to 8 + 8 and 1 more.

8 + 7 = __15__

doubles minus 1

8 + 7 is equal to 1 less than 8 + 8.

Show and Grow — I can do it!

1. Use the double 6 + 6 to find each sum.

6 + 7 = ___

6 + 5 = ___

Name _____

✓ Apply and Grow: Practice

Use the double 7 + 7 to find each sum.

2. 7 + 8 = ___

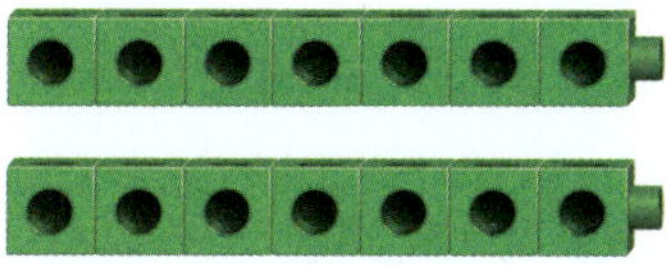

7 + 6 = ___

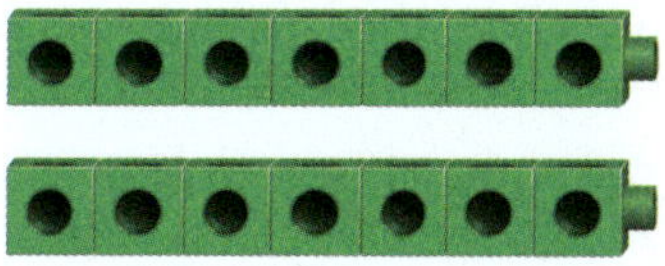

Find the sum. Write the double you used.

3. 5 + 6 = ___

___ ◯ ___ = ___

4. 9 + 8 = ___

___ ◯ ___ = ___

5. **DIG DEEPER!** Circle two ways you can solve 9 + 10. Show how you know.

9 + 9 and 1 more

9 + 9 and 1 less

10 + 10 and 1 more

10 + 10 and 1 less

Think and Grow: Modeling Real Life

A music room has 7 keyboards. There is 1 more recorder than keyboards. How many instruments are there?

Which doubles can you use to find the sum?

7 + 7   8 + 8   6 + 6

Equation:

_____ instruments

Show and Grow   I can think deeper!

6. A museum has 10 paintings. There is 1 fewer sculpture than paintings. How many art pieces are there?

Which doubles can you use to find the sum?

8 + 8   9 + 9   10 + 10

Equation:

_____ art pieces

Name _____

Practice 4.2

Learning Target: Use the *doubles plus 1* and *doubles minus 1* strategies to find a sum.

Use the double 5 + 5 to find each sum.

5 + 6 = **11**

5 + 6 is equal to 5 + 5 and 1 more.

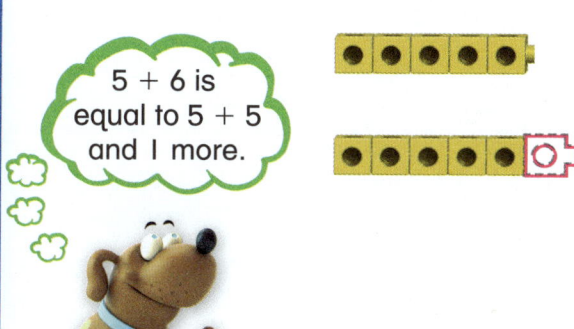

5 + 4 = **9**

5 + 4 is equal to 1 less than 5 + 5.

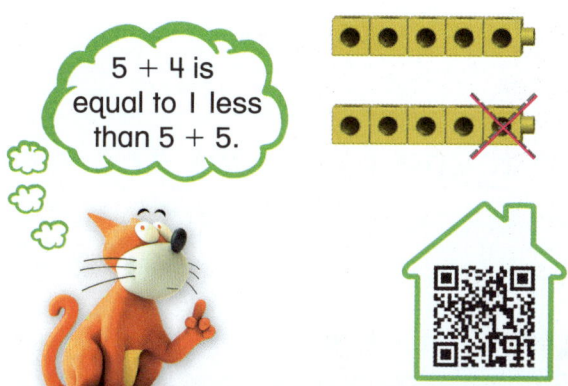

Use the double 9 + 9 to find each sum.

1. 9 + 10 = ____

9 + 8 = ____

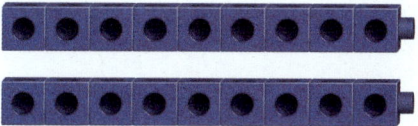

Find the sum. Write the double you used.

2. 10 + 9 = ____

   ___ ◯ ___ = ___

3. 7 + 8 = ____

   ___ ◯ ___ = ___

Chapter 4 | Lesson 2

4. **DIG DEEPER!** Circle two ways you can solve 8 + 9. Show how you know.

   8 + 8 and 1 more

   8 + 8 and 1 less

   9 + 9 and 1 more

   9 + 9 and 1 less

5. **Modeling Real Life** There are 6 balls. There is 1 more toy hoop than balls. How many toys are there?

   Which doubles can you use to find the sum?

   7 + 7   8 + 8   6 + 6

   ____ toys

Review & Refresh

Use the picture to write an equation.

6.

   ____ __ ____ = ____

7.

   ____ __ ____ = ____

Name _____

Count On to Add within 20 4.3

Learning Target: Use the *count on* strategy to find a sum.

Explore and Grow

Model the story.

There are 8 coins in a piggy bank. You put in 5 more. How many coins are in the bank now?

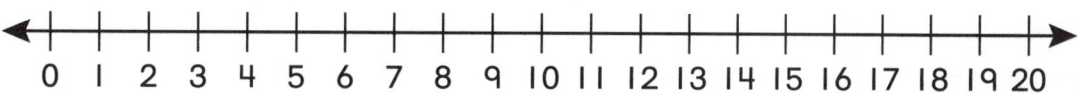

____ coins

Chapter 4 | Lesson 3   one hundred ninety-nine   199

# Think and Grow

One Way:

Start at 7. Count on 9.

$7 + 9 = 16$

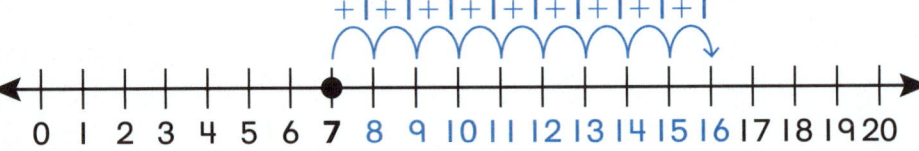

The order of the addends changes, but the sum stays the same!

Another Way:

Start at 9. Count on 7.

$9 + 7 = 16$

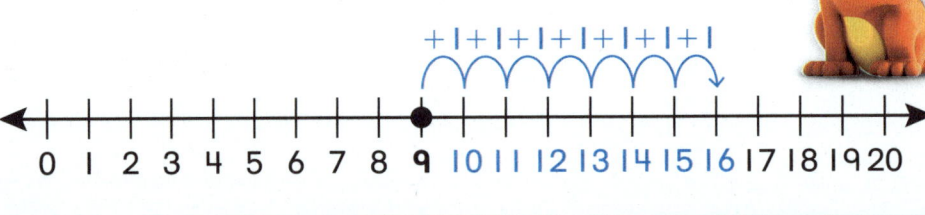

Show and Grow — I can do it!

1. $8 + 4 = \underline{}$

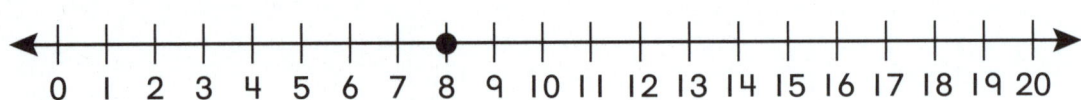

2. $5 + 10 = \underline{}$

Name _____

Apply and Grow: Practice

<----|----|----|----|----|----|----|----|----|----|----|----|----|----|----|----|----|----|----|----|---->
0 1 2 3 4 5 6 7 8 9 10 11 12 13 14 15 16 17 18 19 20

3. 6 + 8 = ____

4. 12 + 5 = ____

5. 7 + 8 = ____

6. 10 + 9 = ____

7. ____ = 0 + 11

8. ____ = 4 + 9

9. **Structure** Write the equation shown by the number line. Then write the equation another way.

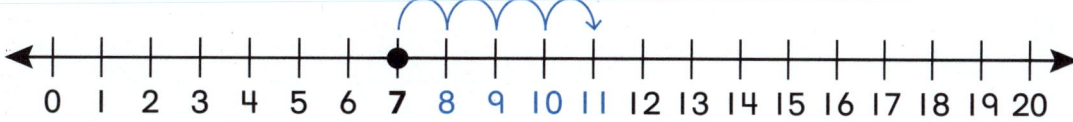

0 1 2 3 4 5 6 **7 8 9 10 11** 12 13 14 15 16 17 18 19 20

____ + ____ = ____   ____ + ____ = ____

Chapter 4 | Lesson 3   two hundred one   201

Think and Grow: Modeling Real Life

You have 7 train cars. Newton has 5 more than you. Descartes has 4 more than you. Who has more train cars, Newton or Descartes?

Model:

Equations: Newton   Descartes

Who has more?   Newton   Descartes

Show and Grow   I can think deeper!

10. You have 6 comic books. Newton has 5 more than you. Descartes has 6 more than you. Who has more comic books, Newton or Descartes?

Model:

Equations: Newton   Descartes

Who has more?   Newton   Descartes

Name _____

Practice  **4.3**

Learning Target: Use the *count on* strategy to find a sum.

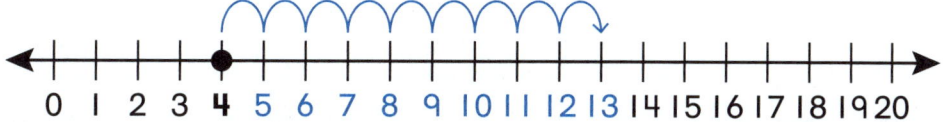

4 + 9 = __13__

1. 10 + 7 = ____

2. 8 + 9 = ____

3. ____ = 6 + 9

4. ____ = 12 + 4

Chapter 4 | Lesson 3   two hundred three   **203**

5. **Structure** Write the equation shown by the number line. Then write the equation another way.

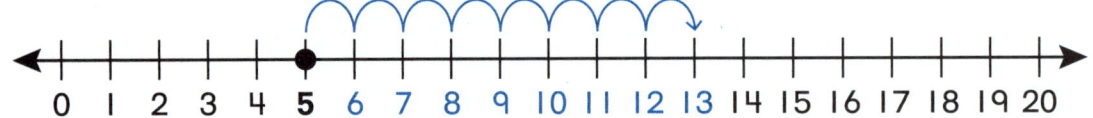

___ + ___ = ___   ___ + ___ = ___

6. **Modeling Real Life** You have 11 toys. Newton has 3 more than you. Descartes has 6 more than you. Who has more toys, Newton or Descartes?

Newton   Descartes

Review & Refresh

7. Circle the triangle. Draw a rectangle around the hexagon.

8. Circle the cube. Draw a rectangle around the sphere.

Name _____

Learning Target: Add three numbers.

Use linking cubes to model the story.

You have 5 red pencils, 4 yellow pencils, and 3 blue pencils. How many pencils do you have in all?

_____ pencils

Chapter 4 | Lesson 4

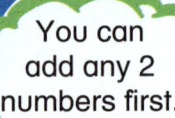 Think and Grow

You can add any 2 numbers first.

4 + 1 + 5 = ?

④+①+ 5 = 10
     |
    [5]

The sum is always the same.

④+ 1 +⑤= 10
    |
   [9]

4 +①+⑤= 10
    |
   [6]

Show and Grow   I can do it!

1.

②+⑤+ 6 = ___
   |
  [ ]

2 +⑤+⑥= ___
    |
   [ ]

②+ 5 +⑥= ___
    |
   [ ]

2.

④+ 3 +④= ___
   |
  [ ]

4 +③+④= ___
    |
   [ ]

④+③+ 4 = ___
    |
   [ ]

206   two hundred six

Name _____

✓ Apply and Grow: Practice

3.

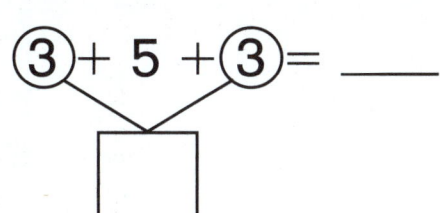 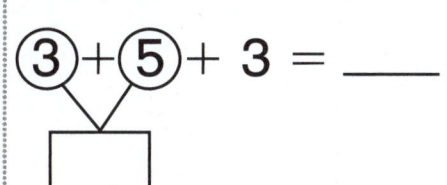 ③ + ⑤ + ③ = ___

4.
3 + 1 + 2 = ___
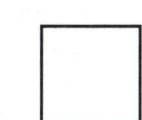

5.
7 + 5 + 5 = ___

6.
4 + 5 + 2 = ___

7.
7 + 8 + 1 = ___

8.
8 + 6 + 2 = ___

9.
4 + 9 + 4 = ___

10. **DIG DEEPER!** Complete the number puzzle so that each branch has a sum of 14.

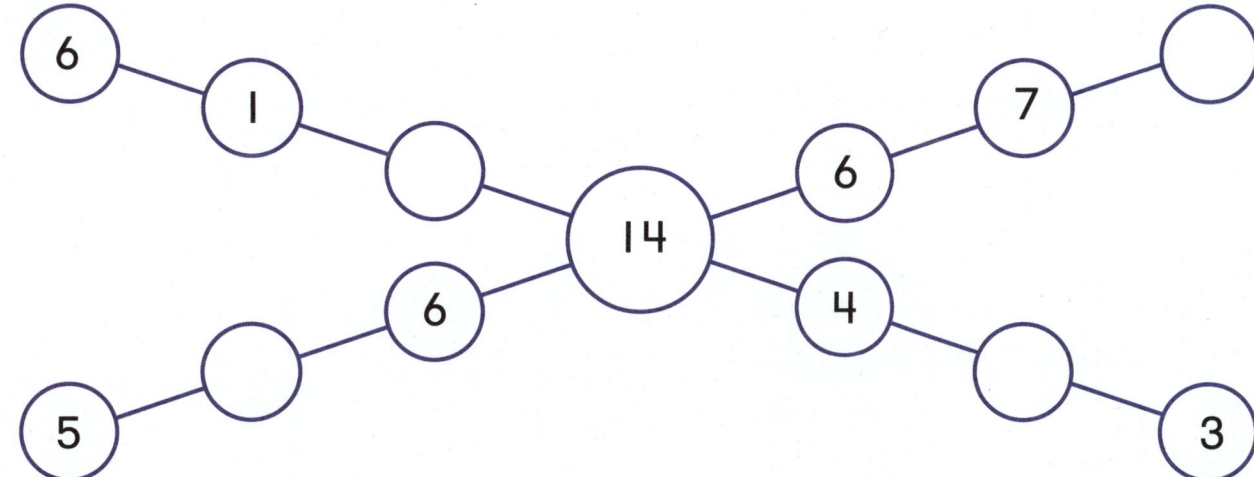

Chapter 4 | Lesson 4   two hundred seven **207**

# Think and Grow: Modeling Real Life

You have 6 goldfish, 7 minnows, and 6 guppies. Will the tank hold all of your fish?

Equation: ___ + ___ + ___ = ___

Yes   No

Show how you know:

Show and Grow   I can think deeper!

11. You have 2 lovebirds, 4 canaries, and 3 finches. Will the cage hold all of your birds?

Equation: ___ + ___ + ___ = ___

Yes   No

Show how you know:

Name _____

Practice 4.4

Learning Target: Add three numbers.

You can add any two numbers first.

6 + 5 + 1 = ?

⑥ + ⑤ + 1 = __12__
            |11|

6 + ⑤ + ① = __12__
           |6|

1.

6 + ② + ② = ____
       | |

⑥ + ② + 2 = ____
       | |

⑥ + 2 + ② = ____
       | |

2.
6 + 5 + 6 = ____
| |

3.
8 + 7 + 4 = ____
| |

4.
9 + 10 + 1 = ____
| |

5.
3 + 4 + 5 = ____

6.
7 + 5 + 5 = ____

7.
8 + 4 + 8 = ____

Chapter 4 | Lesson 4   two hundred nine **209**

8. **DIG DEEPER!** Complete the number puzzle so that each branch has the sum of 13.

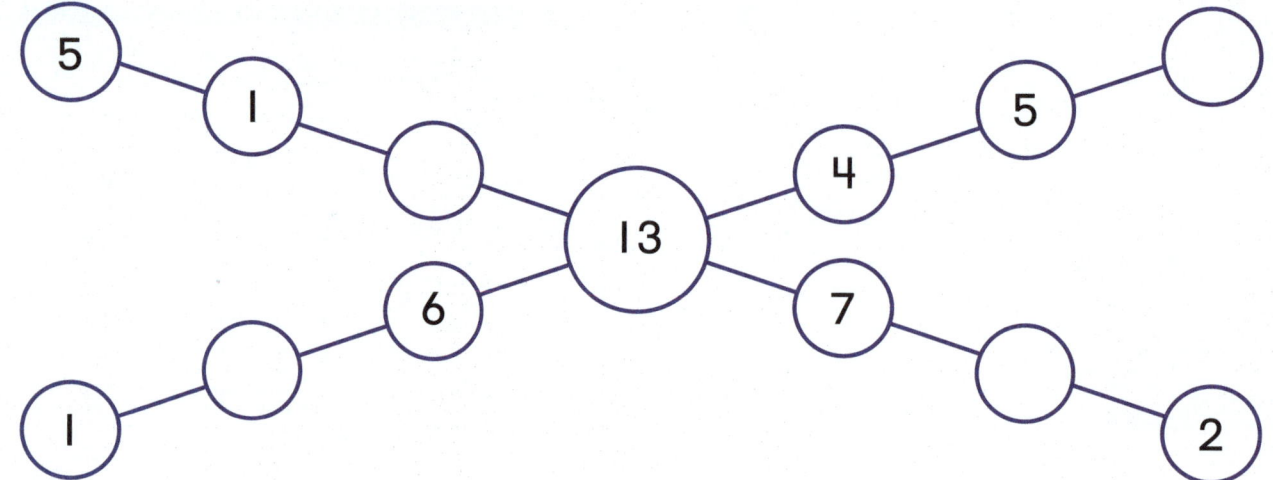

9. **Modeling Real Life** You have 7 white chickens, 1 black chicken, and 8 brown chickens. Will the chicken coop hold all of your chickens?

___ + ___ + ___ = ___

Yes   No

Holds 15 Chickens

Show how you know:

Review & Refresh

10. $5 + \underline{} = 10$

11. $10 = 6 + \underline{}$

12. $8 + \underline{} = 10$

13. $10 = 3 + \underline{}$

Name _____

Learning Target: Use the *make a 10* strategy to add three numbers.

Add Three Numbers by Making a 10   4.5

Explore and Grow

Show three ways to find a sum.

4 + 3 + 7 = ___

4 + 3 + 7 = ___

4 + 3 + 7 = ___

Chapter 4 | Lesson 5   two hundred eleven   211

Think and Grow

Find two addends whose sum is 10. Add those numbers first.

$3 + 7 + 5 = 15$

[10]

```
   8
   4
 + 6   [10]
 ----
  18
```

Making a 10 can help you add three numbers.

Show and Grow — I can do it!

Make a 10 to add.

1. $9 + 1 + 3 = $ ____

2. $4 + 2 + 8 = $ ____

[10] [10]

3.
```
    2
    3
  + 7    [10]
  ---
  [  ]
```

4.
```
    5
    5
  + 9    [10]
  ---
  [  ]
```

212　two hundred twelve

Name _____

✓ Apply and Grow: Practice

Make a 10 to add.

5. $3 + 7 + 3 =$ ___
   [10]

6. $8 + 6 + 2 =$ ___
   [10]

7.
   ```
     6
     7    [10]
   + 4
   ─────
    [ ]
   ```

8.
   ```
     8
     5    [10]
   + 5
   ─────
    [ ]
   ```

9. $4 + 6 + 2 =$ ___

10. $5 + 9 + 1 =$ ___

11. **DIG DEEPER!** What do you know about the missing addends and the sum?

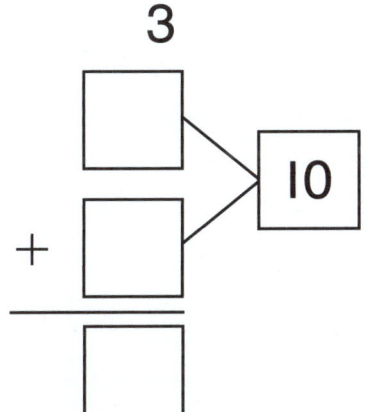

Chapter 4 | Lesson 5   two hundred thirteen 213

Think and Grow: Modeling Real Life

You need to knock down 20 pins to win. Do you win?

Equation: ___ + ___ + ___ = ___

Pins
6
3
7

Yes   No

Show how you know:

Show and Grow — I can think deeper!

12. Your hockey team needs 12 goals to break a record. Does your team break the record?

    Equation: ___ + ___ + ___ = ___

Goals
5
6
4

    Yes   No

    Show how you know:

Name _____

Practice 4.5

Learning Target: Use the *make a 10* strategy to add three numbers.

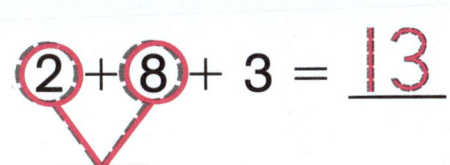

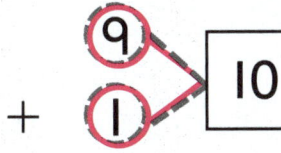

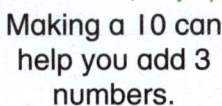

Making a 10 can help you add 3 numbers.

Find 2 addends whose sum is 10. Add those numbers first.

Make a 10 to add.

1. 6 + 8 + 4 = ___   10

2.   3
       7   10
   +   4
   ☐

3. 5 + 8 + 2 = ___

4. 1 + 9 + 7 = ___

Chapter 4 | Lesson 5   two hundred fifteen **215**

5. **DIG DEEPER!** What do you know about the missing addends and the sum?

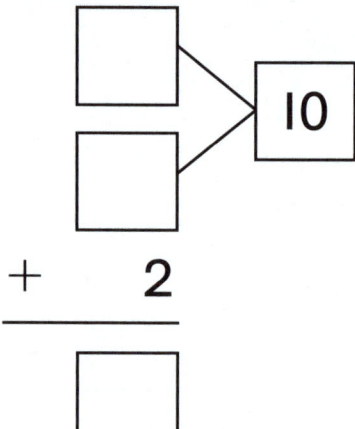

6. **Modeling Real Life** Your baseball team needs 15 runs to break a record. Does your team break the record?

___ + ___ + ___ = ___

Yes   No

Runs
6
4
6

Show how you know:

Review & Refresh

Circle the heavier object.

7.

8.

Name _____

Learning Target: Use the *make a 10* strategy when adding 9.

Add 9 **4.6**

Explore and Grow

Use counters and the ten frames to find the sum. Show how you can make a 10 to solve.

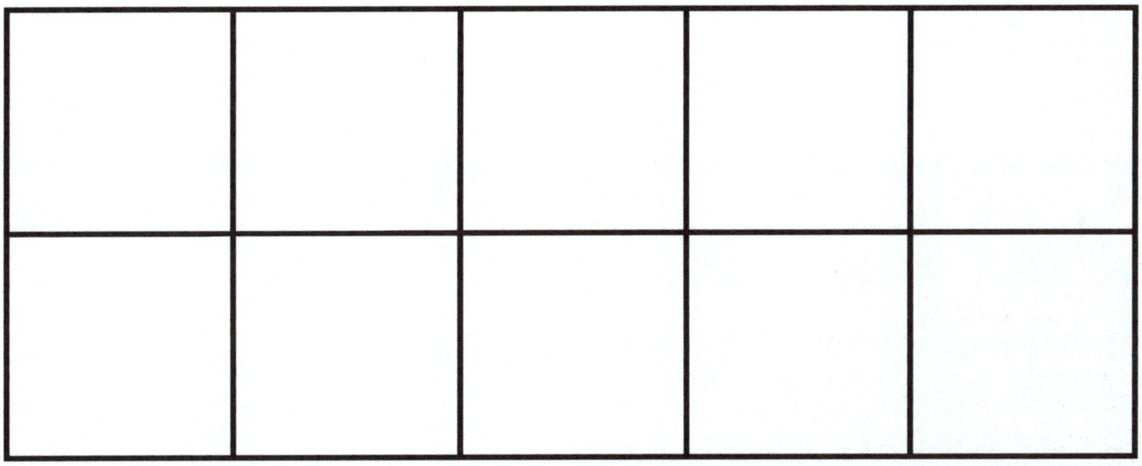

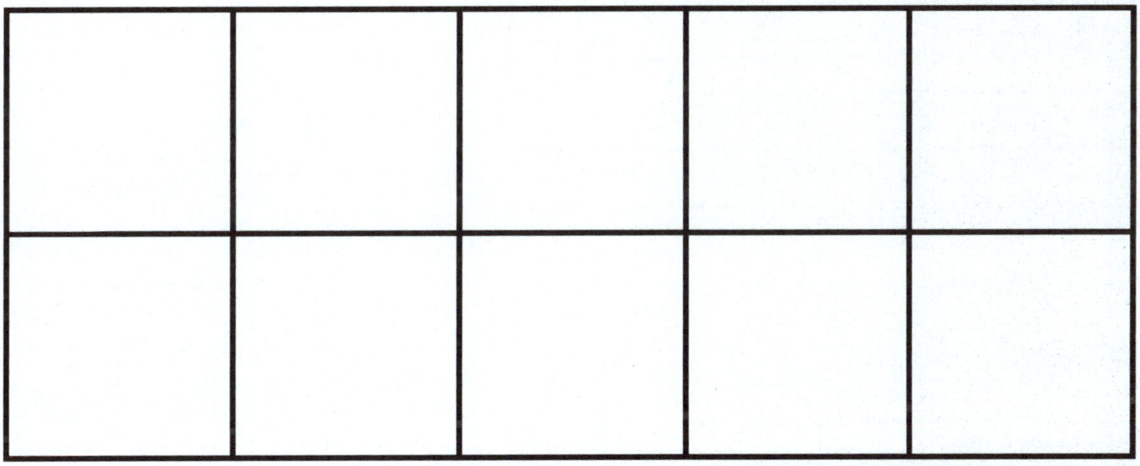

9 + 5 = ___

Chapter 4 | Lesson 6

Think and Grow

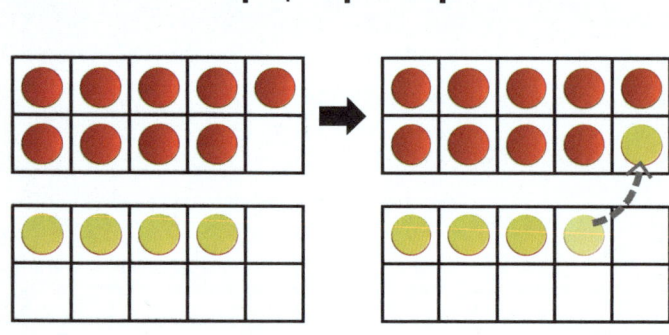

$9 + 4 = ?$

$9 + 4$
$9 + \underline{1} + \underline{3}$
$10 + \underline{3} = \underline{13}$
So, $9 + 4 = \underline{13}$.

Show and Grow I can do it!

Make a 10 to add.

1. $9 + 3 = ?$

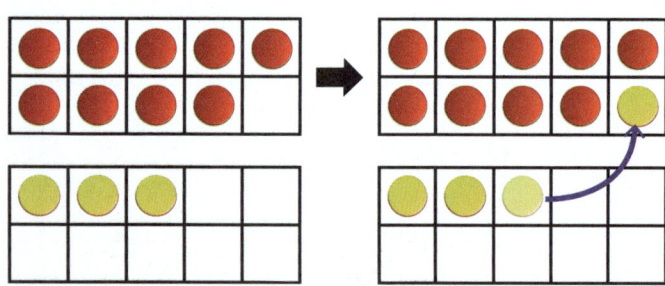

$9 + 3$
$9 + \underline{} + \underline{}$
$10 + \underline{} = \underline{}$
So, $9 + 3 = \underline{}$.

2. $9 + 6 = ?$

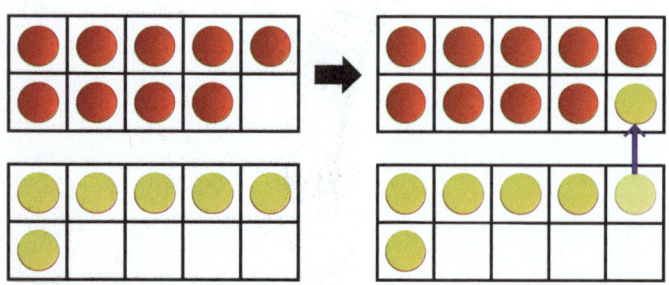

$9 + 6$
$9 + \underline{} + \underline{}$
$10 + \underline{} = \underline{}$
So, $9 + 6 = \underline{}$.

Apply and Grow: Practice

Make a 10 to add.

3. 9 + 2 = ?

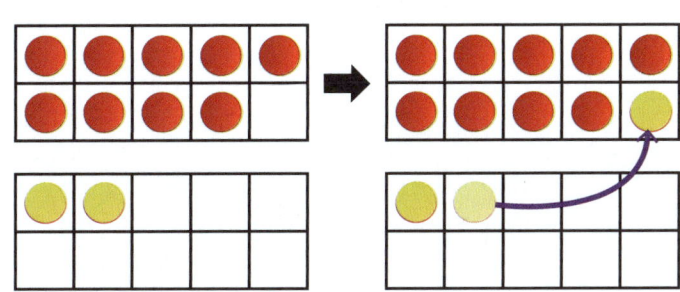

9 + 2

9 + ___ + ___

10 + ___ = ___

So, 9 + 2 = ___.

4. 9 + 9 = ?

9 + 9

9 + ___ + ___

10 + ___ = ___

So, 9 + 9 = ___.

5. 9 + 5 = ?

9 + 5

9 + ___ + ___

10 + ___ = ___

So, 9 + 5 = ___.

6. **DIG DEEPER!** Use the ten frame to complete the equations.

9 + ? = 17

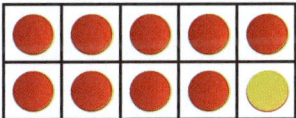

10 + ___ = 17

So, 9 + ___ = 17.

Chapter 4 | Lesson 6

two hundred nineteen 219

Think and Grow: Modeling Real Life

You have 9 stickers and earn 5 more. Your friend has 6 stickers and earns 10 more. Do you and your friend have the same number of stickers?

Addition equations: You   Friend

Yes   No

Show how you know:

Show and Grow   I can think deeper!

7. Your friend has 9 magnets and finds 6 more. You have 5 magnets and find 10 more. Do you and your friend have the same number of magnets?

Addition equations: You   Friend

Yes   No

Show how you know:

Name _____

Practice  4.6

Learning Target: Use the *make a 10* strategy when adding 9.

$9 + 5 = ?$

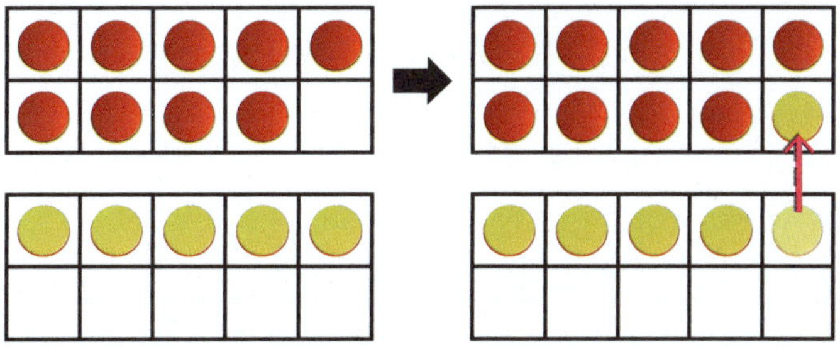

Make a 10 by thinking of 5 as 1 + 4.

$9 + 5$
$9 + \underline{1} + \underline{4}$
$10 + \underline{4} = \underline{14}$
So, $9 + 5 = \underline{14}$.

Make a 10 to add.

1. $9 + 3 = ?$

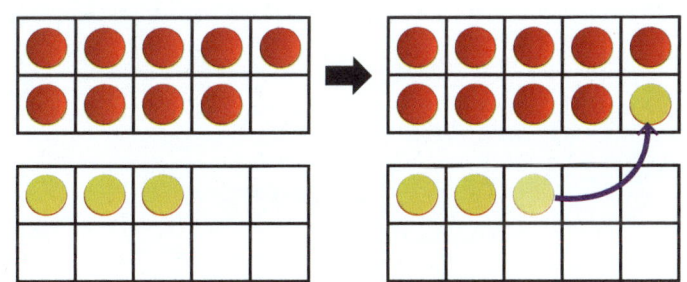

$9 + 3$
$9 + \underline{} + \underline{}$
$10 + \underline{} = \underline{}$
So, $9 + 3 = \underline{}$.

Chapter 4 | Lesson 6   two hundred twenty-one 221

2. **DIG DEEPER!** Use the ten frame to complete the equations.

   16 = ? + ?

   10 + ___ = ___

   So, 16 = ___ + ___.

3. **Modeling Real Life** You have 9 rocks and collect 7 more. Your friend has 8 rocks and collects 8 more. Do you and your friend have the same number of rocks?

   Yes   No

   Show how you know:

   Review & Refresh

   Circle the longer object.

   4.

   5.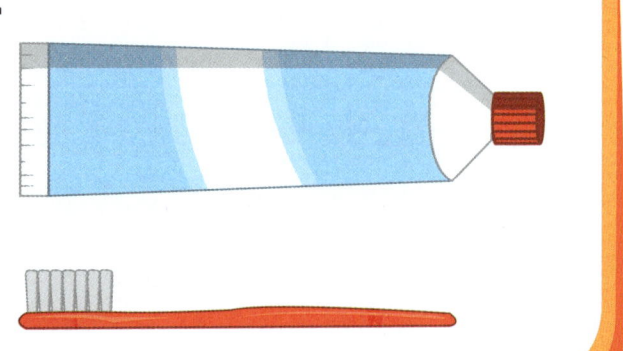

   222   two hundred twenty-two

Name _____

Make a 10 to Add 4.7

Learning Target: Use the *make a 10* strategy to add two numbers.

Explore and Grow

Use counters and the ten frame to find the sum. Show how you can make a 10 to solve.

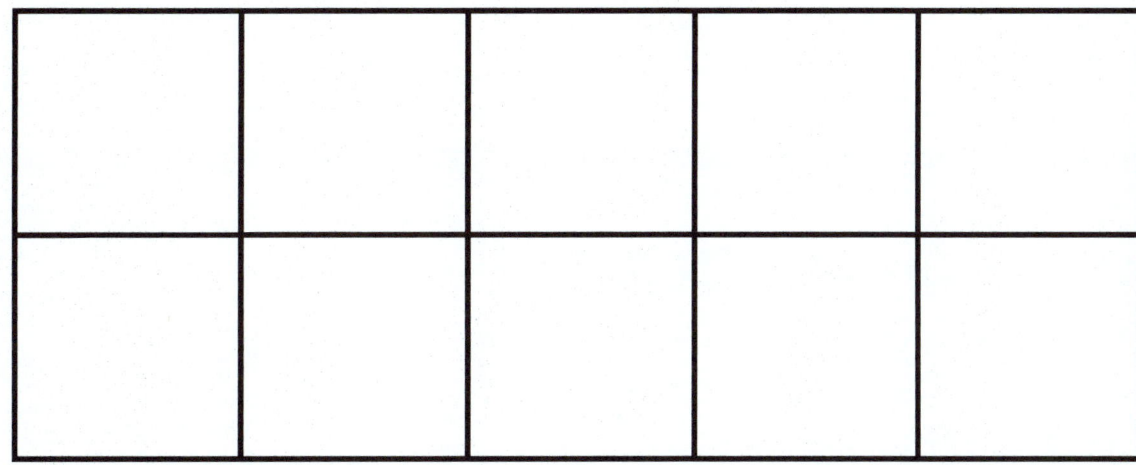

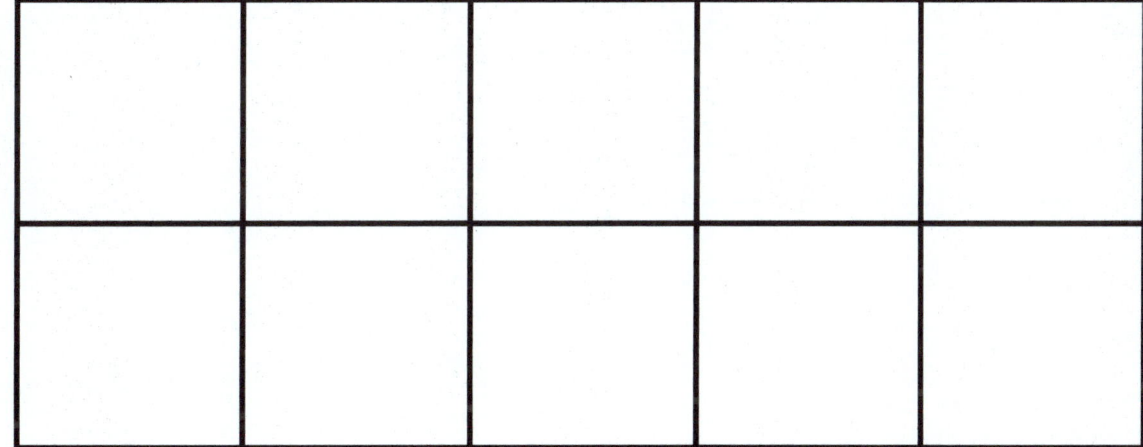

8 + 6 = ___

Chapter 4 | Lesson 7

# Think and Grow

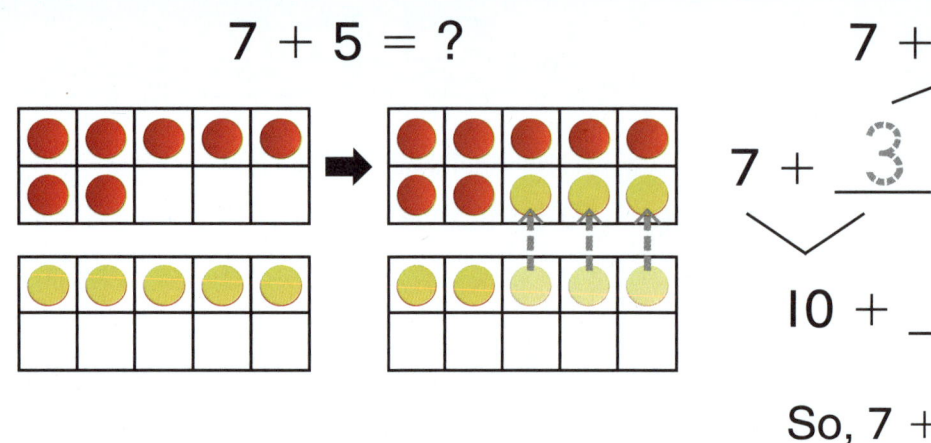

$7 + 5 = ?$

$7 + 5$

$7 + \underline{3} + \underline{2}$

$10 + \underline{2} = \underline{12}$

So, $7 + 5 = \underline{12}$.

Show and Grow — I can do it!

Make a 10 to add.

1. $8 + 5 = ?$

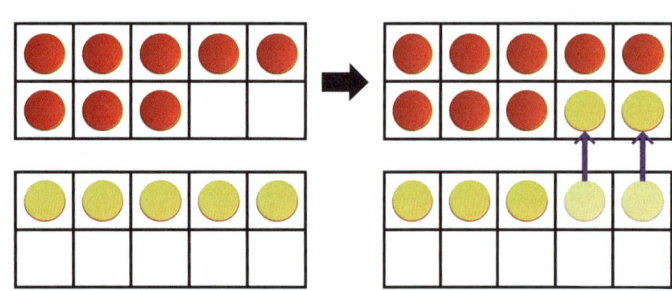

$8 + 5$

$8 + \underline{} + \underline{}$

$10 + \underline{} = \underline{}$

So, $8 + 5 = \underline{}$.

2. $7 + 7 = ?$

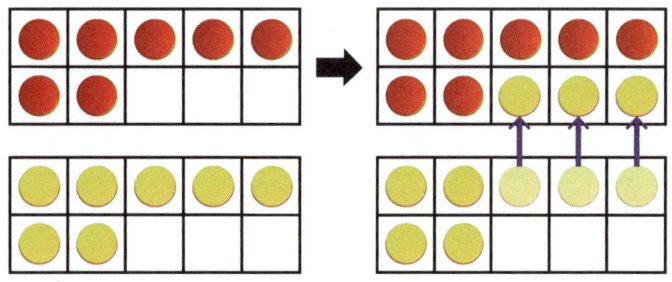

$7 + 7$

$7 + \underline{} + \underline{}$

$10 + \underline{} = \underline{}$

So, $7 + 7 = \underline{}$.

Name _____

Apply and Grow: Practice

Make a 10 to add.

3. 7 + 6 = ?

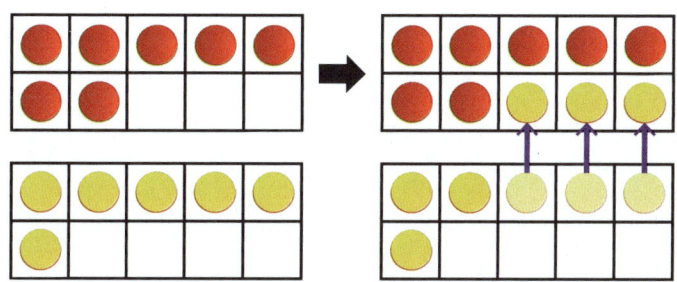

7 + 6
7 + ___ + ___
10 + ___ = ___

So, 7 + 6 = ___.

4. 7 + 4 = ?

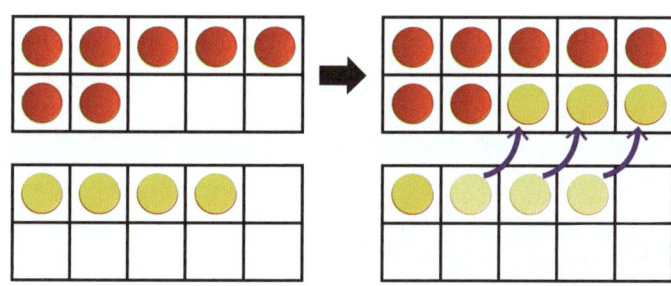

7 + 4
7 + ___ + ___
10 + ___ = ___

So, 7 + 4 = ___.

5.   8 + 7
8 + ___ + ___
10 + ___ = ___

So, 8 + 7 = ___.

6.   6 + 6
6 + ___ + ___
10 + ___ = ___

So, 6 + 6 = ___.

7. **Number Sense** Use 4, 7, and 10 to complete the sentence.

7 + ___ has the same sum as ___ + ___.

Chapter 4 | Lesson 7   two hundred twenty-five   **225**

Think and Grow: Modeling Real Life

There are 8 crabs. 7 more join them. There are 10 turtles. 5 more join them. Is the number of crabs the same as the number of turtles?

Addition equations:

Yes   No

Show how you know:

Show and Grow — I can think deeper!

8. There are 6 squirrels. 6 more join them. There are 10 birds. 2 more join them. Is the number of squirrels the same as the number of birds?

Addition equations:

Yes   No

Show how you know:

Name _____

Practice

Learning Target: Use the *make a 10* strategy to add two numbers.

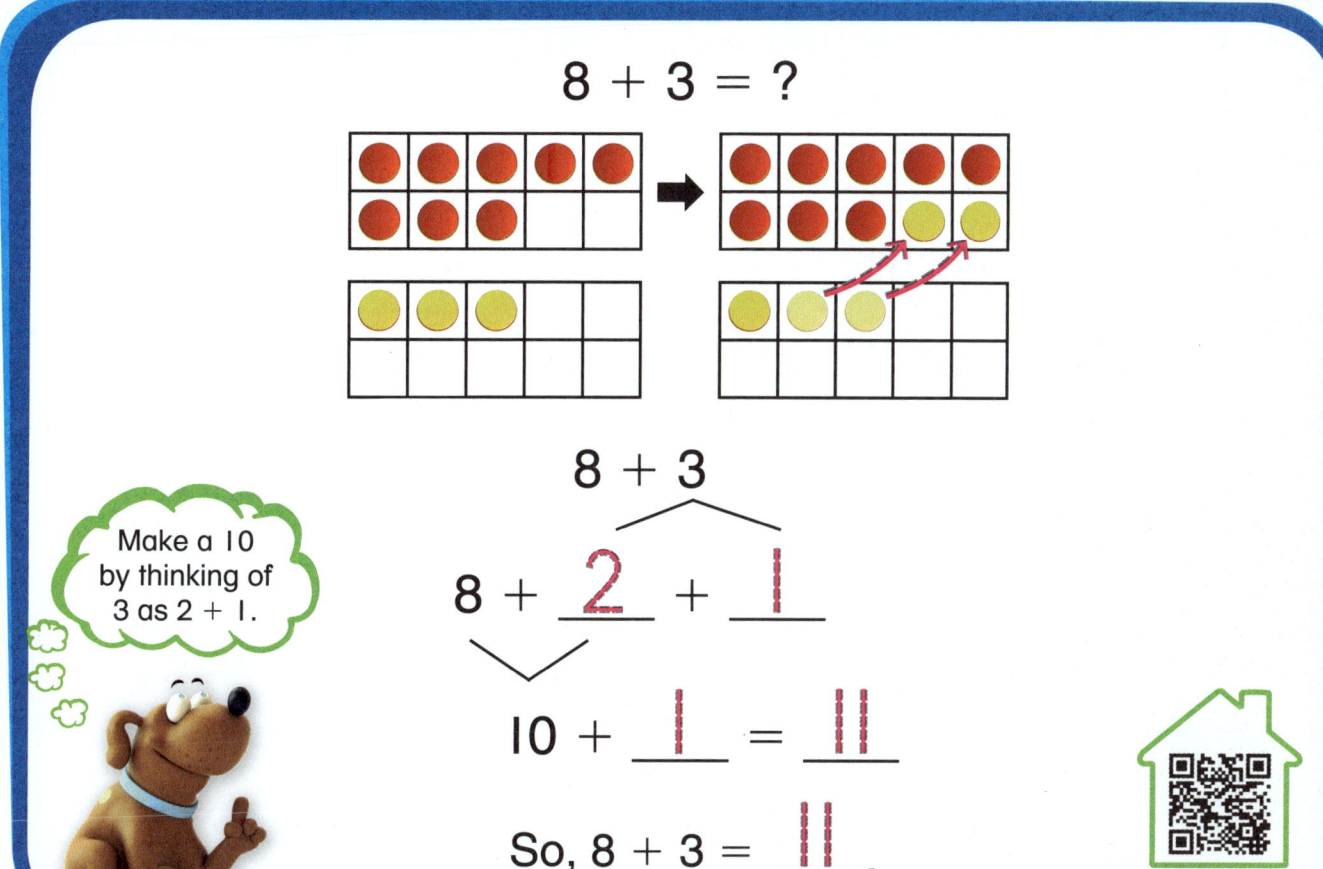

Make a 10 to add.

1. 6 + 5 = ?

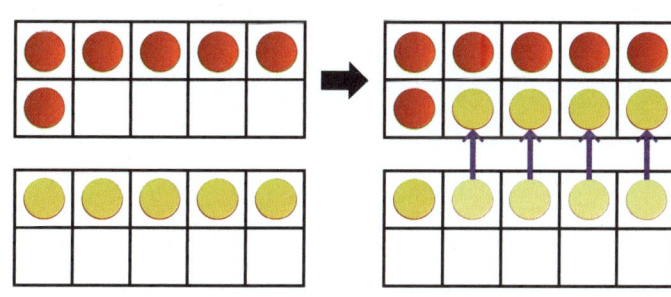

6 + 5

6 + ___ + ___

10 + ___ = ___

So, 6 + 5 = ___.

Chapter 4 | Lesson 7   two hundred twenty-seven   227

2. **Number Sense** Use 2, 6, and 10 to complete the sentence.

    ____ + ____ has the same sum as ____ + 6.

3. **Modeling Real Life** There are 7 monkeys. 4 more join them. There are 6 birds. 5 more join them. Is the number of monkeys the same as the number of birds?

    Yes   No

    Show how you know:

Review & Refresh

4. There are 3 ✈.

   2 more ✈ join them.

   How many ✈ are there now?

   ____ + ____ = ____

   ____

5. There are 4 🦈.

   6 more 🦈 join them.

   How many 🦈 are there now?

   ____ + ____ = ____

   ____

228   two hundred twenty-eight

Name _____

Learning Target: Solve addition word problems.

Problem Solving: Addition within 20 — 4.8

Explore and Grow

Newton has 9 red crayons, 3 blue crayons, and 1 yellow crayon. How many crayons does he have in all?

_____ crayons

Chapter 4 | Lesson 8

Think and Grow

There are 8 girls and 6 boys in your class.

How many students are in your class?

Circle what you know.

Underline what you need to find.

Solve:

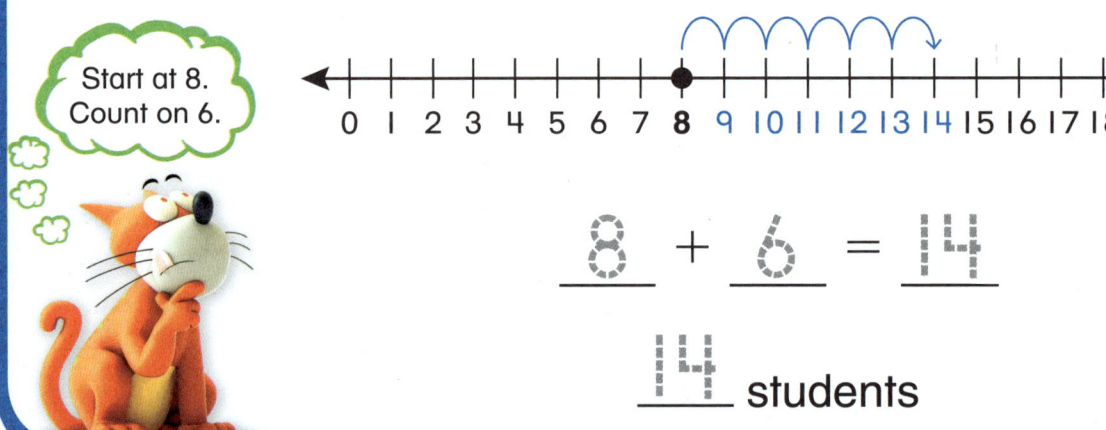

$\underline{8} + \underline{6} = \underline{14}$

$\underline{14}$ students

Show and Grow — I can do it!

1. You have 6 notebooks. You buy 5 more. How many notebooks do you have now?

   Circle what you know.

   Underline what you need to find.

   Solve:

   ___ ◯ ___ = ___   ___ notebooks

Apply and Grow: Practice

2. There are 6 soccer balls, 10 basketballs, and 4 volleyballs. How many balls are there in all?

   Circle what you know.   Underline what you need to find.

   Solve:

   ___ ◯ ___ ◯ ___ = ___

   ☐

   ___ balls

3. You do 8 push-ups. Your friend does 1 fewer than you. How many push-ups do you and your friend do in all?

   ___ push-ups

4. **YOU BE THE TEACHER** Newton has 9 magnets. Descartes has 8 more than Newton. Your friend uses a bar model to show how many magnets Descartes has. Is your friend correct? Show how you know.

   Newton: | 9 |

   Descartes: | 8 | 1 |

   $8 + 1 = 9$

   9 magnets

Chapter 4 | Lesson 8

Think and Grow: Modeling Real Life

You have 5 bracelets. You have 7 fewer than your friend. How many bracelets does your friend have?

Circle what you know.

Underline what you need to find.

Solve: Friend: [   ]
         You: [   ]

___ + ___ = ___   ___ bracelets

Show and Grow   I can think deeper!

5. Your friend finds 9 seashells. You find 6 more than your friend. How many seashells do you find?

   Circle what you know.

   Underline what you need to find.

   Solve:

   ___ + ___ = ___   ___ seashells

Name _____

Practice 4.8

Learning Target: Solve addition word problems.

You have 6 keys. Your friend has 1 more than you. How many keys do you and your friend have in all?

Circle what you know.   Underline what you need to find.

6 + 7 is equal to 6 + 6 and 1 more.

Solve:

__6__ ⊕ __7__ = __13__

__13__ keys

1. You have 5 robots. Your friend gives you more. Now you have 14. How many robots did your friend give you?

   Circle what you know.   Underline what you need to find.

   Solve:

   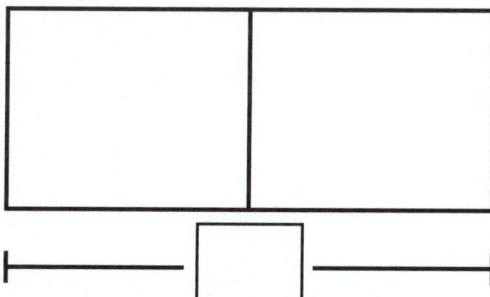

   ___ ◯ ___ = ___

   ___ robots

2. There are some plates on a table. You add 8 more. Now there are 12. How many plates were on the table to start?

   ___ plates

Chapter 4 | Lesson 8

3. **YOU BE THE TEACHER** Newton has 8 tickets. Descartes has 5 more than Newton. Your friend uses a number line to show how many tickets Descartes has. Is your friend correct? Show how you know.

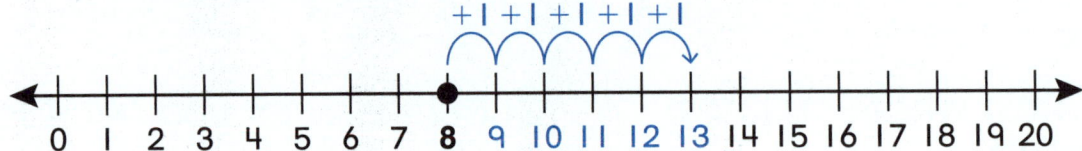

13 tickets

4. **Modeling Real Life** You have 9 medals. You have 9 fewer than your friend. How many medals does your friend have?

Friend:
You:

_____ medals

Review & Refresh

5. 4 − 3 = _____

6. 5 − 2 = _____

7. 9 − 9 = _____

8. 7 − 0 = _____

Name _____

Performance Task 4

1. You track the weather for a few weeks. Each week has 7 days.

   a. You track the weather every day for the first week. But you miss 1 day in the second week. How many days do you track the weather?

   _____ days

   b. You track the weather for 1 more week. How many days in all do you track the weather?

   _____ days

2. Your friend also tracks the weather. She records 9 sunny days and 5 cloudy days. How many days does your friend track the weather?

   _____ days

3. You record 10 rainy days in the first three weeks. Is the number of rainy days the same as the number of sunny days?

   Yes   No

Week	Sunny Days
1	4
2	3
3	4

   Show how you know:

Chapter 4   two hundred thirty-five   235

Roll and Cover

To Play: Roll 3 dice and find the sum. Place a counter on a fish with the sum. Take turns until all of the fish have been covered.

Name _____

Chapter 4 Practice

4.1 Add Doubles from 6 to 10

1.

___ + ___ = ___

2.

4.2 Use Doubles within 20

Use the double 6 + 6 to find each sum.

3. 6 + 7 = ___

6 + 5 = ___

Find the sum. Write the double you used.

4. 7 + 8 = ___

___ ◯ ___ = ___

5. 10 + 9 = ___

___ ◯ ___ = ___

Chapter 4   two hundred thirty-seven   237

4.3 Count On to Add within 20

6. $11 + 6 =$ _____

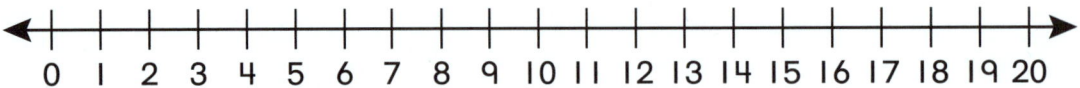

4.4 Add Three Numbers

7.

$6 + ④ + ④ =$ _____ | $⑥ + ④ + 4 =$ _____ | $⑥ + 4 + ④ =$ _____

8. $7 + 8 + 5 =$ _____

9. $9 + 8 + 1 =$ _____

10. $2 + 3 + 5 =$ _____

238 two hundred thirty-eight

 4.5 Add Three Numbers by Making a 10

Make a 10 to add.

11. $7 + 9 + 3 =$ _____

12. $4 + 9 + 1 =$ _____

[10]

[10]

13. **Number Sense** What do you know about the missing addends and the sum?

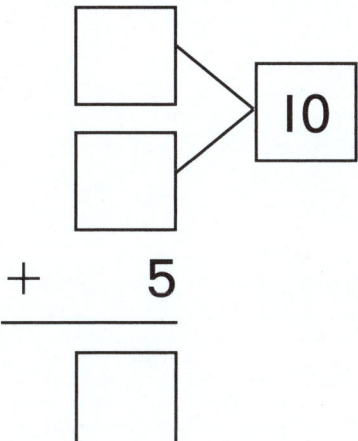

 4.6 Add 9

14. Make a 10 to add $9 + 4$.

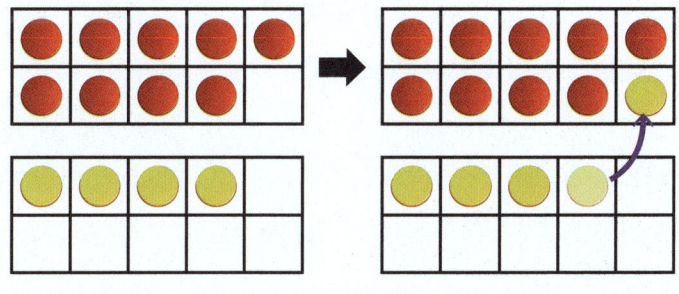

$9 + 4$

$9 + $ ___ $ + $ ___

$10 + $ ___ $ = $ ___

So, $9 + 4 = $ ___.

Chapter 4   two hundred thirty-nine   **239**

4.7 Make a 10 to Add

15. Make a 10 to add 8 + 5.

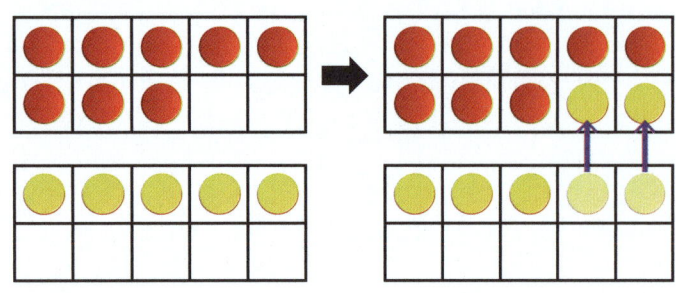

8 + 5

8 + ___ + ___

10 + ___ = ___

So, 8 + 5 = ___.

16. 7 + 7 = ?

7 + 7

7 + ___ + ___

10 + ___ = ___

So, 7 + 7 = ___.

17. 9 + 8 = ?

9 + 8

9 + ___ + ___

10 + ___ = ___

So, 9 + 8 = ___.

4.8 Problem Solving: Addition within 20

18. Modeling Real Life Your friend finds 7 insects. You find 9 more than your friend. How many insects do you find?

___ insects

5 Subtract Numbers within 20

- What do bees make?
- How many bees do you see? 7 of them fly away. How many bees are left?

Chapter Learning Target:
Understand subtraction strategies.

Chapter Success Criteria:
- I can identify counting back strategies.
- I can describe subtraction equations.
- I can explain the subtraction strategy I used.
- I can compare addition and subtraction strategies.

5 Vocabulary

Organize It

Review Words
bar model
difference
minus
part-part-whole model
subtraction equation

Use the review words to complete the graphic organizer.

8 − 3 = 5

Define It

Match the review word to its definition.

1. bar model

2. part-part-whole model

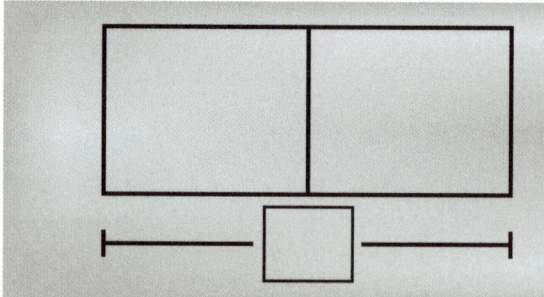

Name _____

Count Back to Subtract within 20 — 5.1

Learning Target: Use the *count back* strategy to find a difference.

Explore and Grow

Model the story.

You are on floor number 12. You go down 4 floors. What floor are you on now?

floor number ____

Chapter 5 | Lesson 1   two hundred forty-three   243

Think and Grow

Start at 13. Count back 5.

13 − 5 = __8__

Show and Grow   I can do it!

1. 15 − 8 = ____

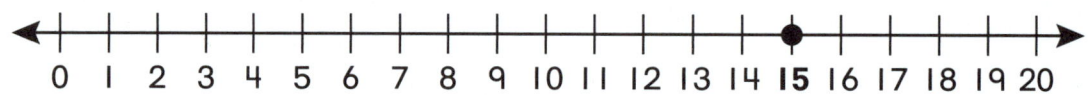

2. 12 − 3 = ____

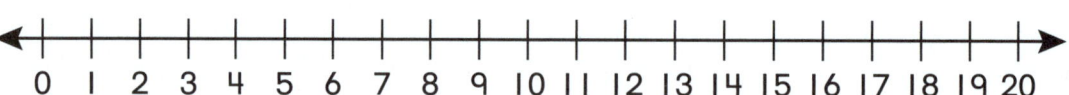

3. 17 − 9 = ____

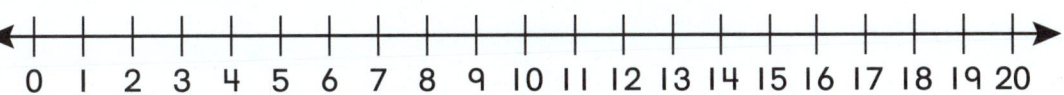

Apply and Grow: Practice

4. 11 − 6 = ___

5. 13 − 7 = ___

6. 18 − 9 = ___

7. 20 − 10 = ___

8. 17 − 8 = ___

9. 18 − 3 = ___

10. ___ = 16 − 4

11. ___ = 14 − 8

12. **DIG DEEPER!** Write the equation shown by the number line.

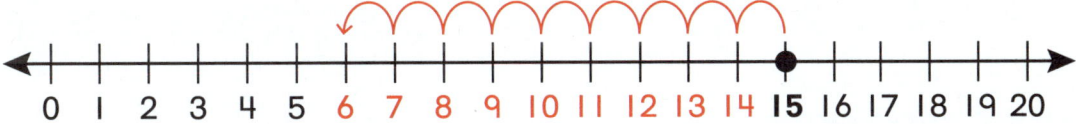

___ − ___ = ___

Chapter 5 | Lesson 1 two hundred forty-five **245**

# Think and Grow: Modeling Real Life

You collect 6 gems. Your friend collects 14. How many fewer gems do you collect?

Model:

Subtraction equation:

____ fewer gems

Show and Grow   I can think deeper!

13. Your friend finds 16 gold bars. You find 9. How many fewer gold bars do you find?

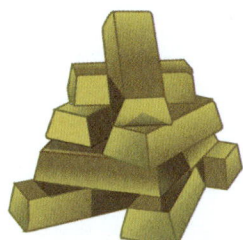

Model:

Subtraction equation:

____ gold bars

Name _____

Practice 5.1

Learning Target: Use the *count back* strategy to find a difference.

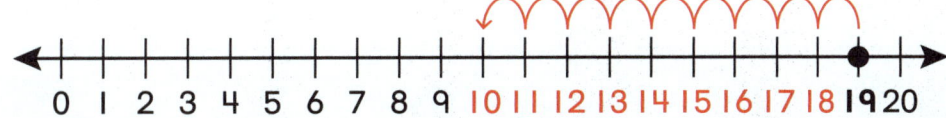

$19 - 9 = \underline{10}$

1. $12 - 5 = \underline{}$

2. $15 - 7 = \underline{}$

3. $13 - 6 = \underline{}$

4. $14 - 5 = \underline{}$

5. $11 - 5 = \underline{}$

6. $15 - 4 = \underline{}$

7. $\underline{} = 18 - 8$

8. $\underline{} = 12 - 8$

Chapter 5 | Lesson 1

two hundred forty-seven 247

9. **DIG DEEPER!** Write the equation shown by the number line.

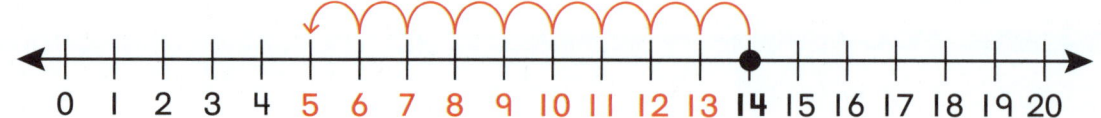

___ − ___ = ___

10. **Modeling Real Life** You play soccer. The visiting team scores 12 goals. Your team scores 4 fewer. How many goals does your team score?

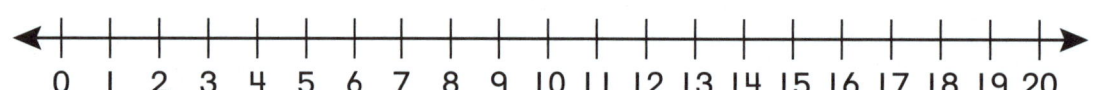

___ goals

Review & Refresh

11. $6 + 2 =$ ___

12. $4 + 3 =$ ___

13. $8 + 2 =$ ___

14. $5 + 4 =$ ___

Name _____

Learning Target: Use the *add to subtract* strategy to find a difference.

Explore and Grow

Model the story.

Your class needs to make 15 scrapbook pages. 8 are already made. How many more pages does your class need to make?

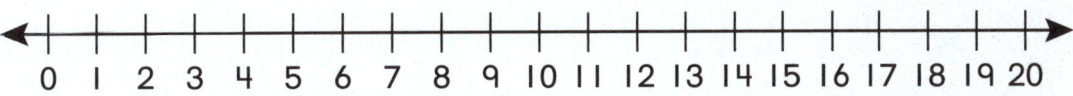

_____ more pages

Chapter 5 | Lesson 2

Think and Grow

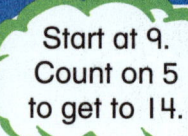

Start at 9. Count on 5 to get to 14.

$14 - 9 = ?$

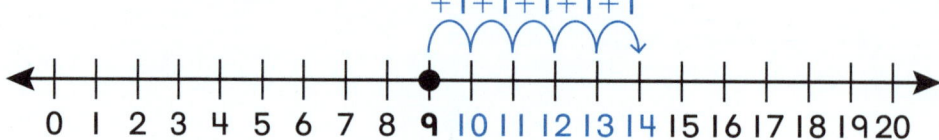

Think $9 + \underline{5} = 14$.

So, $14 - 9 = \underline{5}$.

Show and Grow — I can do it!

1. $11 - 7 = ?$

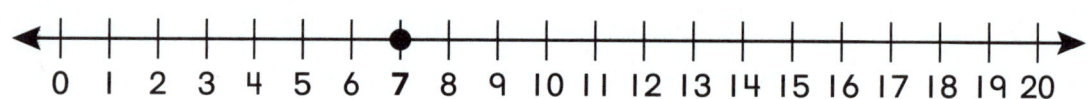

Think $7 + \underline{} = 11$.   So, $11 - 7 = \underline{}$.

2. $16 - 8 = ?$

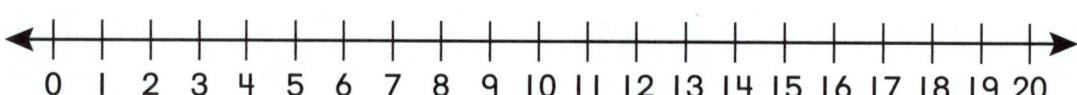

Think $8 + \underline{} = 16$.   So, $16 - 8 = \underline{}$.

3. $13 - 10 = ?$

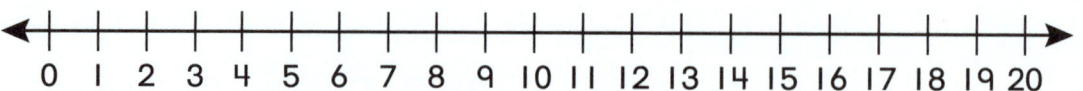

Think $10 + \underline{} = 13$.   So, $13 - 10 = \underline{}$.

Apply and Grow: Practice

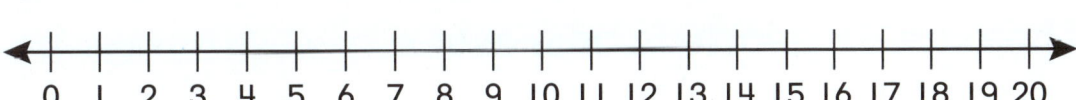

4. 12 − 7 = ?

   Think 7 + ___ = 12.

   So, 12 − 7 = ___.

5. 17 − 8 = ?

   Think 8 + ___ = 17.

   So, 17 − 8 = ___.

6. 11 − 2 = ___

7. ___ = 19 − 5

8. **DIG DEEPER!** Tell what subtraction problem Newton and Descartes solved. Think: What strategies did they use?

 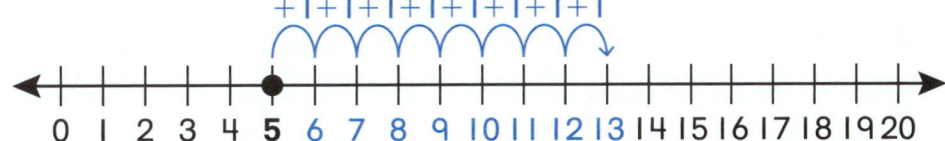

 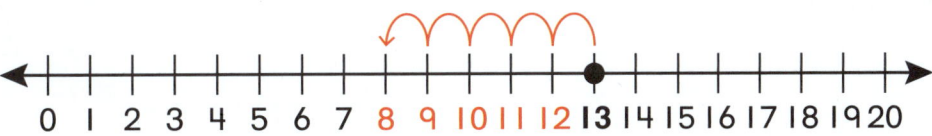

___ − ___ = ___

Chapter 5 | Lesson 2

Think and Grow: Modeling Real Life

There are 15 people in an elevator. Some of them exit. There are 7 left. How many people exit?

Model:

Subtraction equation:

_____ people

Show and Grow   I can think deeper!

9. There are 18 people in a subway car. Some of them exit. There are 9 left. How many people exit?

   Model:

   Subtraction equation:

   _____ people

Name _____

Practice 5.2

Learning Target: Use the *add to subtract* strategy to find a difference.

12 − 5 = ?

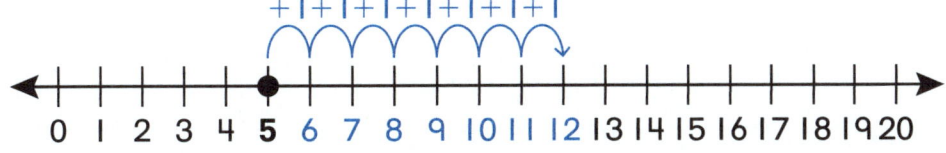

Think 5 + __7__ = 12.

So, 12 − 5 = __7__.

1. 13 − 9 = ?

   Think 9 + ___ = 13.

   So, 13 − 9 = ___.

2. 17 − 8 = ?

   Think 8 + ___ = 17.

   So, 17 − 8 = ___.

3. 14 − 7 = ___

4. ___ = 17 − 4

Chapter 5 | Lesson 2

two hundred fifty-three 253

5. **DIG DEEPER!** Tell what subtraction problem Newton and Descartes solved. Think: What strategies did they use?

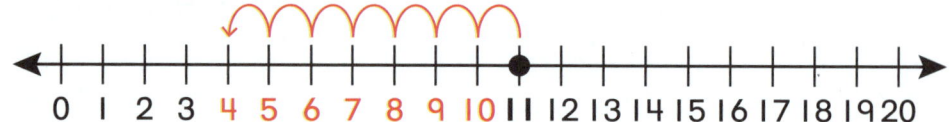

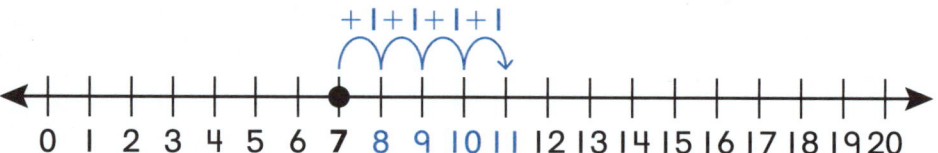

___ − ___ = ___

6. **Modeling Real Life** There are 13 people on a train. Some of them exit. There are 6 left. How many people exit?

___ people

Review & Refresh

7. 10 − 9 = ___

8. 9 − 9 = ___

Name _____

Subtract 9 5.3

Learning Target: Use the *get to 10* strategy when subtracting 9.

Explore and Grow

Use counters to find each difference.

15 − 10 = ___   15 − 9 = ___

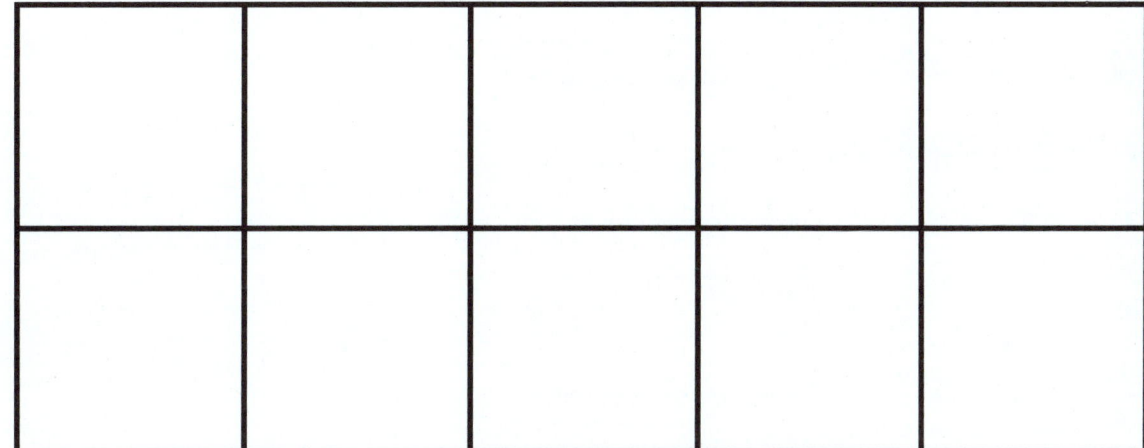

Chapter 5 | Lesson 3

Think and Grow

Start at 13. Subtract 3 to get to 10. 9 = 3 + 6, so subtract 6 more.

13 − 9 = ?

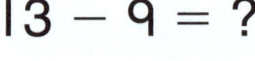

13 − __3__ = 10

10 − __6__ = __4__

So, 13 − 9 = __4__.

Show and Grow — I can do it!

Get to 10 to subtract.

1. 17 − 9 = ?

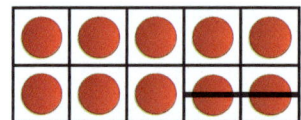

17 − ___ = 10

10 − ___ = ___

So, 17 − 9 = ___.

2. 14 − 9 = ?

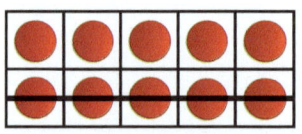

14 − ___ = 10

10 − ___ = ___

So, 14 − 9 = ___.

256   two hundred fifty-six

Name _____

✓ Apply and Grow: Practice

3. 16 − 9 = ?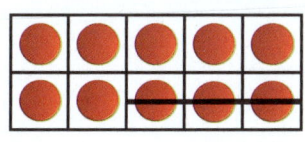

16 − ___ = 10

10 − ___ = ___

So, 16 − 9 = ___.

4. 11 − 9 = ?

11 − ___ = 10

10 − ___ = ___

So, 11 − 9 = ___.

5. 15 − 9 = ?

15 − ___ = 10

10 − ___ = ___

So, 15 − 9 = ___.

6. **MP Structure** Which models show 12 − 9?

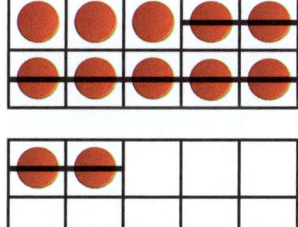 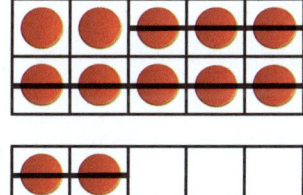 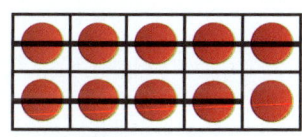
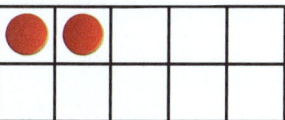

Think and Grow: Modeling Real Life

You have 12 eggs. You use 9 of them. How many eggs are left?

Model: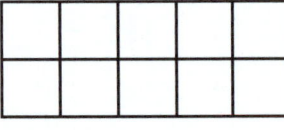

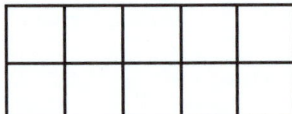

Subtraction equation:

_____ eggs

Show and Grow I can think deeper!

7. An egg carton has 18 eggs. You crack 9 of them. How many eggs are *not* cracked?

    Model: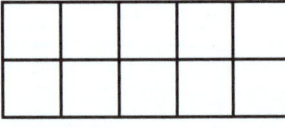

    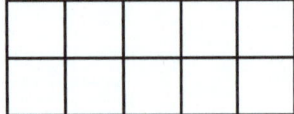

    Subtraction equation:

_____ eggs

Name _____

Practice 5.3

Learning Target: Use the *get to 10* strategy when subtracting 9.

15 − 9 = ?

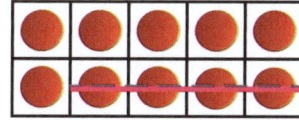

15 − __5__ = 10

10 − __4__ = __6__

So, 15 − 9 = __6__.

Get to 10 to subtract.

1. 18 − 9 = ?

   18 − ___ = 10

   10 − ___ = ___

   So, 18 − 9 = ___.

2. 12 − 9 = ?

   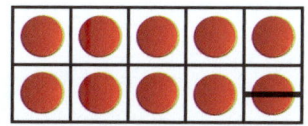

   12 − ___ = 10

   10 − ___ = ___

   So, 12 − 9 = ___.

3. 19 − 9 = ?

   19 − ___ = 10

   10 − ___ = ___

   So, 19 − 9 = ___.

Chapter 5 | Lesson 3   two hundred fifty-nine **259**

4. **Structure** Which models show 16 − 9?

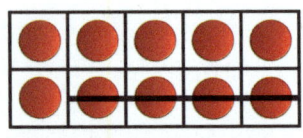
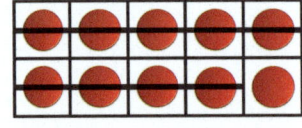
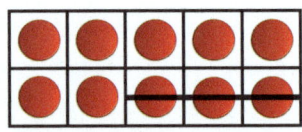

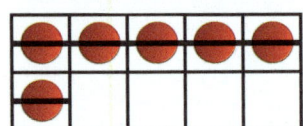
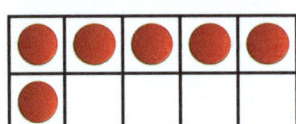
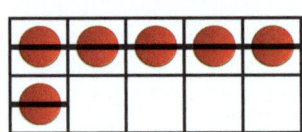

5. **Modeling Real Life** You have 14 water balloons. You break 9 of them. How many water balloons are left?

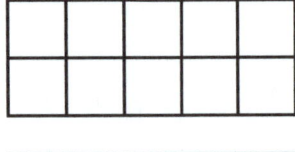

_____ water balloons

Review & Refresh

6. Use the picture to complete the number bond.

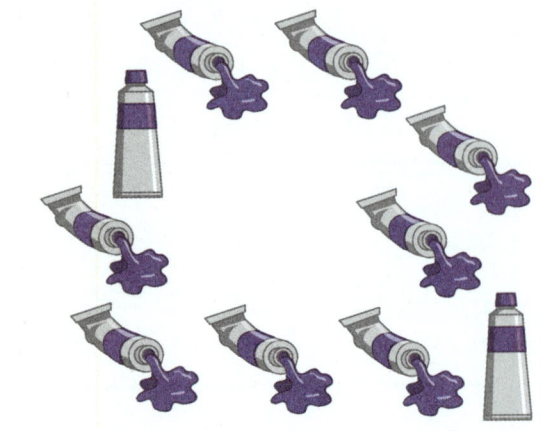
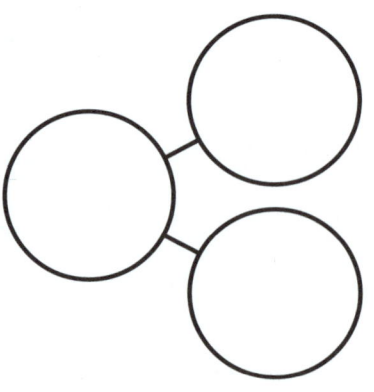

Name _____

Learning Target: Use the *get to 10* strategy to subtract.

Explore and Grow

Use counters to find the difference. Show how you can make a 10 to solve.

$$14 - 6 = \underline{}$$

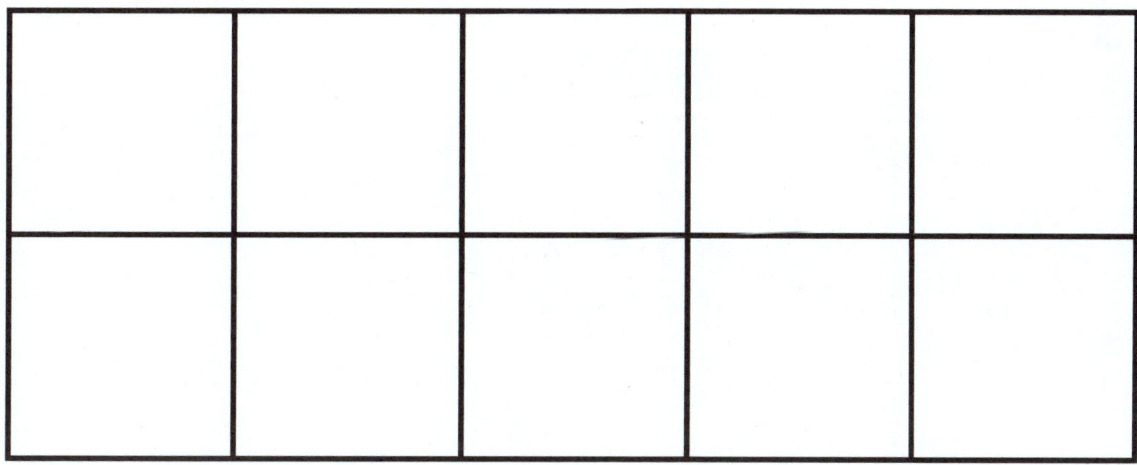

Chapter 5 | Lesson 4

Think and Grow

Start at 15. Subtract 5 to make a 10. 7 = 5 + 2, so subtract 2 more.

15 − 7 = ?

15 − 5 = 10

10 − 2 = 8

So, 15 − 7 = 8.

Show and Grow I can do it!

Get to 10 to subtract.

1. 12 − 5 = ?

   12 − ___ = 10

   10 − ___ = ___

   So, 12 − 5 = ___.

2. 17 − 8 = ?

   17 − ___ = 10

   10 − ___ = ___

   So, 17 − 8 = ___.

Name _____

Apply and Grow: Practice

Get to 10 to subtract.

3. 16 − 7 = ?

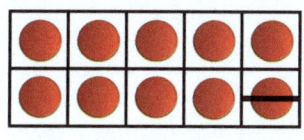

16 − ___ = 10

10 − ___ = ___

So, 16 − 7 = ___.

4. 11 − 4 = ?

11 − ___ = 10

10 − ___ = ___

So, 11 − 4 = ___.

5. 13 − 5 = ?

13 − ___ = 10

10 − ___ = ___

So, 13 − 5 = ___.

6. **Number Sense** Which equations did Newton use to solve the problem?

○ 17 − 7 = 10,   10 − 7 = 3

○ 17 − 2 = 15,   15 − 4 = 11

○ 12 − 2 = 10,   10 − 4 = 6

Chapter 5 | Lesson 4

Think and Grow: Modeling Real Life

Your friend checks out 13 books. You check out 4 fewer. How many books do you check out?

Model:

Subtraction equation:

_____ books

Show and Grow   I can think deeper!

7. Your friend skips 16 stones. You skip 8 fewer. How many stones do you skip?

Model:

Subtraction equation:

_____ stones

Name _____

Practice 5.4

Learning Target: Use the *get to 10* strategy to subtract.

11 − 5 = ?

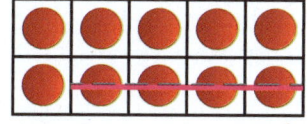

11 − __1__ = 10

10 − __4__ = __6__

So, 11 − 5 = __6__.

Get to 10 to subtract.

1. 15 − 6 = ?

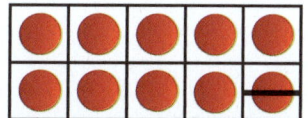

15 − ___ = 10

10 − ___ = ___

So, 15 − 6 = ___.

2. 12 − 4 = ?

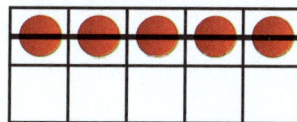

12 − ___ = 10

10 − ___ = ___

So, 12 − 4 = ___.

3. 13 − 7 = ?

13 − ___ = 10

10 − ___ = ___

So, 13 − 7 = ___.

4. 14 − 8 = ?

14 − ___ = 10

10 − ___ = ___

So, 14 − 8 = ___.

Chapter 5 | Lesson 4

two hundred sixty-five 265

5. **Number Sense** Which equations did Descartes use to solve the problem?

○ 15 − 5 = 10, 10 − 4 = 6

○ 20 − 5 = 15, 15 − 5 = 10

○ 20 − 4 = 16, 16 − 5 = 11

6. **Modeling Real Life** Your friend recycles 14 cans. You recycle 7 fewer. How many cans do you recycle?

_____ cans

Review & Refresh

Is the equation true or false?

7. $3 + 2 \stackrel{?}{=} 5 + 0$

   $\underline{} \stackrel{?}{=} \underline{}$

   True   False

8. $8 − 2 \stackrel{?}{=} 5 + 5$

   $\underline{} \stackrel{?}{=} \underline{}$

   True   False

Name _____

Learning Target: Identify whether an equation is true or false.

More True or False Equations 5.5

Explore and Grow

Color the stars that have a sum or difference equal to 19 − 5.

- 9 + 5
- 20 − 6
- 12 + 1 + 1
- 16 − 2
- 13 − 1
- 14 + 0

Chapter 5 | Lesson 5

 **Think and Grow**

$$15 - 8 \overset{?}{=} 11 - 4$$

15 − 8:

11 − 4:

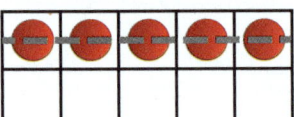
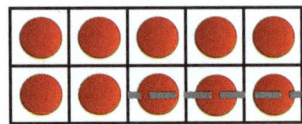
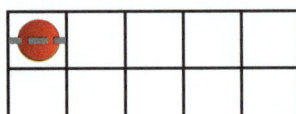

$$\underline{}7 \overset{?}{=} \underline{}7$$

(True)   False

Show and Grow — I can do it!

Is the equation true or false?

1. $17 - 9 \overset{?}{=} 14 - 5$ 17 − 9:   14 − 5:

   $\underline{} \overset{?}{=} \underline{}$

   True   False

2. $6 + 5 \overset{?}{=} 18 - 7$ 6 + 5:   18 − 7:

   $\underline{} \overset{?}{=} \underline{}$

   True   False

Name _____

✓ Apply and Grow: Practice

Is the equation true or false?

3. $5 + 7 \stackrel{?}{=} 3 + 8$ $5 + 7$: ___ $3 + 8$: ___

___ $\stackrel{?}{=}$ ___

True   False

4. $4 + 9 \stackrel{?}{=} 5 + 3 + 5$ $4 + 9$: ___ $5 + 3 + 5$: ___

___ $\stackrel{?}{=}$ ___

True   False

5. $12 - 7 \stackrel{?}{=} 13 - 5$

___ $\stackrel{?}{=}$ ___

True   False

6. $14 - 8 \stackrel{?}{=} 12 - 6$

___ $\stackrel{?}{=}$ ___

True   False

7. $1 + 8 \stackrel{?}{=} 16 - 7$

___ $\stackrel{?}{=}$ ___

True   False

8. $18 - 9 \stackrel{?}{=} 5 + 1 + 4$

___ $\stackrel{?}{=}$ ___

True   False

9. **Number Sense** Circle all of the equations that are true.

$20 \stackrel{?}{=} 2$ $19 - 7 \stackrel{?}{=} 12$ $6 + 6 + 3 \stackrel{?}{=} 7 + 8$

Chapter 5 | Lesson 5   two hundred sixty-nine   **269**

Think and Grow: Modeling Real Life

You have 12 lemons. You use 4 of them. Your friend has 3 lemons and buys 4 more. Do you each have the same number of lemons?

Equation: ___ − ___ $\stackrel{?}{=}$ ___ + ___

___ $\stackrel{?}{=}$ ___

Yes   No

Show and Grow   I can think deeper!

10. You have 14 grapes. You eat 7 of them. Your friend has 10 grapes and eats 3 of them. Do you each have the same number of grapes?

Equation: ___ − ___ $\stackrel{?}{=}$ ___ − ___

___ $\stackrel{?}{=}$ ___

Yes   No

Name _____

Practice 5.5

Learning Target: Identify whether an equation is true or false.

$$5 + 9 \stackrel{?}{=} 7 + 8$$

5 + 9: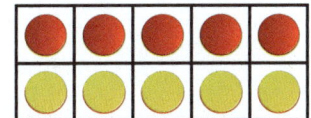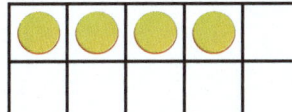

7 + 8: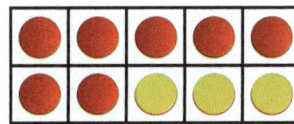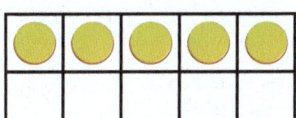

$$14 \stackrel{?}{=} 15$$

True (**False**)

Is the equation true or false?

1. $13 - 9 \stackrel{?}{=} 11 - 7$ 13 − 9: ___   11 − 7: ___

   ___ $\stackrel{?}{=}$ ___

   True   False

2. $9 + 8 \stackrel{?}{=} 17 + 0$ True

   ___ $\stackrel{?}{=}$ ___   False

3. $12 - 4 \stackrel{?}{=} 15 - 6$ True

   ___ $\stackrel{?}{=}$ ___   False

4. $4 + 5 \stackrel{?}{=} 11 - 3$ True

   ___ $\stackrel{?}{=}$ ___   False

5. $15 - 7 \stackrel{?}{=} 4 + 4$ True

   ___ $\stackrel{?}{=}$ ___   False

Is the equation true or false?

6. $0 + 5 + 2 \stackrel{?}{=} 12 - 5$

   ___ $\stackrel{?}{=}$ ___

   True　　False

7. $16 - 8 \stackrel{?}{=} 2 + 3 + 2$

   ___ $\stackrel{?}{=}$ ___

   True　　False

8. **Number Sense** Circle all of the equations that are false.

   $7 + 2 \stackrel{?}{=} 11 - 2$　　$3 \stackrel{?}{=} 12 - 8$　　$4 + 1 + 4 \stackrel{?}{=} 14 - 6$

9. **Modeling Real Life** You have 9 badges. You earn 3 more. Your friend has 5 badges and earns 7 more. Do you each have the same number of badges?

   ___ + ___ $\stackrel{?}{=}$ ___ + ___

   ___ $\stackrel{?}{=}$ ___

   Yes　　No

Review & Refresh

10. ___ + 2 = 6

11. ___ + 3 = 9

Name _____

Learning Target: Find the number that makes an equation true.

Make True Equations 5.6

Complete the equation.

$$14 - 5 = 3 + \underline{}$$

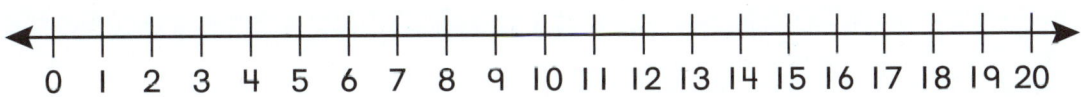

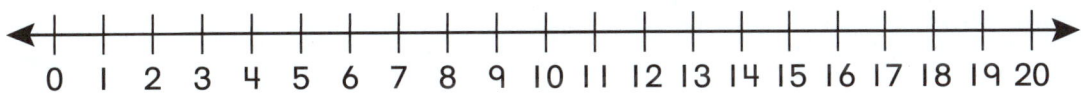

Chapter 5 | Lesson 6

Think and Grow

$13 - ? = 4 + 2$

I know the value of the right side of the equation. $4 + 2 = 6$

$13 - ? = \underline{6}$

The left side has to equal 6, and $13 - 7 = 6$.

$13 - \underline{7} = \underline{6}$

So, $13 - \underline{7} = 4 + 2$.

Show and Grow — I can do it!

1. $? + 6 = 10 - 3$

   $? + 6 = \underline{}$

   $\underline{} + 6 = \underline{}$

   So, $\underline{} + 6 = 10 - 3$.

2. $12 - 9 = ? - 5$

   $\underline{} = ? - 5$

   $\underline{} = \underline{} - 5$

   So, $12 - 9 = \underline{} - 5$.

Name _____

✓ Apply and Grow: Practice

3. $? + 8 = 9 + 6$

   $? + 8 = $ ___

   ___ $+ 8 = $ ___

   So, ___ $+ 8 = 9 + 6$.

4. $13 - 8 = ? - 6$

   ___ $= ? - 6$

   ___ $= $ ___ $- 6$

   So, $13 - 8 = $ ___ $- 6$.

5. $14 - 6 = ? + 2$

   ___ $= ? + 2$

   ___ $= $ ___ $+ 2$

   So, $14 - 6 = $ ___ $+ 2$.

6. $15 - ? = 3 + 3$

   $15 - ? = $ ___

   $15 - $ ___ $= $ ___

   So, $15 - $ ___ $= 3 + 3$.

7. ___ $- 6 = 3 + 2 + 1$

8. $6 + 4 + 4 = 6 + $ ___

9. **YOU BE THE TEACHER** Newton says 2 makes the equation true. Is Newton correct? Show how you know.

$7 + 2 = 11 - ?$

Chapter 5 | Lesson 6

Think and Grow: Modeling Real Life

You catch 11 butterflies. 4 fly away. Your friend catches 3 butterflies. How many more butterflies must your friend catch to have the same number as you?

Equation:

___ − ___ = ___ + ?

___ butterflies

Show and Grow — I can think deeper!

10. You catch 15 leaves. 6 of them blow away. Your friend catches 12 leaves and some of them blow away. Now you each have the same number of leaves. How many of your friend's leaves blow away?

Equation:

___ − ___ = ___ − ?

___ leaves

Name _____

Practice 5.6

Learning Target: Find the number that makes an equation true.

I know the value of the left side of the equation. $5 + 3 = 8$

$5 + 3 = ? - 4$

$\underline{\ 8\ } = ? - 4$

$\underline{\ 8\ } = \underline{\ 12\ } - 4$

So, $5 + 3 = \underline{\ 12\ } - 4$.

1. $12 - ? = 8 - 4$

$12 - ? = \underline{\ \ \ }$

$12 - \underline{\ \ \ } = \underline{\ \ \ }$

So, $12 - \underline{\ \ \ } = 8 - 4$.

2. $7 + 9 = 8 + ?$

$\underline{\ \ \ } = 8 + ?$

$\underline{\ \ \ } = 8 + \underline{\ \ \ }$

So, $7 + 9 = 8 + \underline{\ \ \ }$.

3. $4 + 3 = \underline{\ \ \ } - 7$

4. $7 + \underline{\ \ \ } = 17 - 0$

5. $\underline{\ \ \ } + 20 = 8 + 10 + 2$

6. $3 + 1 + 1 = 14 - \underline{\ \ \ }$

Chapter 5 | Lesson 6

two hundred seventy-seven 277

7. **YOU BE THE TEACHER** Descartes says 5 makes the equation true. Is Descartes correct? Show how you know.

$17 - 9 = ? - 2$

8. **Modeling Real Life** You catch 14 fireflies. You lose 8 of them. Your friend catches 11 fireflies and loses some of them. Now you each have the same number of fireflies. How many fireflies does your friend lose?

_____ − _____ = _____ − ?

_____ fireflies

Review & Refresh

9. Color the rectangles.

   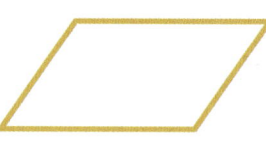

10. Color the squares.

   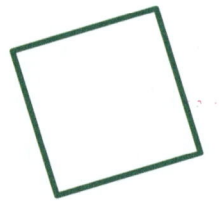

Problem Solving: Subtraction within 20

Learning Target: Solve subtraction word problems.

Explore and Grow

Model the story.

There are 18 seagulls. Some of them fly away. There are 9 left. How many seagulls flew away?

_____ seagulls

Think and Grow

(There are 14 kids in a bounce house.) Some of them get out. (There are 5 kids left.) How many kids got out of the bounce house?

Circle what you know.

Underline what you need to find.

Solve:

Start at 14. Count back to 5.

$14 - 9 = 5$

___9___ kids

Show and Grow I can do it!

1. You have some stuffed animals. You give 3 away. You have 8 left. How many stuffed animals did you have to start?

   Circle what you know. Underline what you need to find.

   Solve:

   ___ ◯ ___ = ___

   ____ stuffed animals

Name _____

✓ Apply and Grow: Practice

2. A group of students are at an arcade. 8 of them leave. There are 3 left. How many students were at the arcade to start?

Circle what you know.   Underline what you need to find.

Solve:

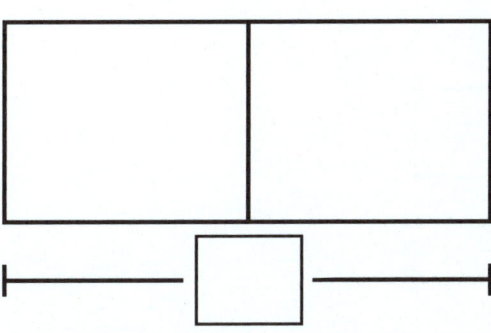

___ ◯ ___ = ___

___ students

3. You have 15 trucks. Your friend has 7. How many more trucks do you have?

___ more trucks

4. **DIG DEEPER!** You have 16 stickers. Your friend has 7 fewer than you. Which bar model shows how many stickers your friend has?

You: | 16 |
Friend: | 7 | 9 |

16 − 9 = 7

You: | 16 |
Friend: | 9 | 7 |

16 − 7 = 9

Chapter 5 | Lesson 7

Think and Grow: Modeling Real Life

Your friend's mask has 13 feathers. Your mask has 7 feathers. How many fewer feathers does your mask have?

Circle what you know.   Underline what you need to find.

Solve:

Friend:

You:

____ − ____ = ____

____ fewer feathers

Show and Grow   I can think deeper!

5. There are 12 party hats. There are 5 fewer noisemakers than party hats. How many noisemakers are there?

Circle what you know.   Underline what you need to find.

Solve:

____ − ____ = ____

____ noisemakers

Name _____

Practice 5.7

Learning Target: Solve subtraction word problems.

You have 11 pebbles. You toss some of them.
You have 4 left. How many pebbles did you toss?

Circle what you know.   Underline what you need to find.

Solve:

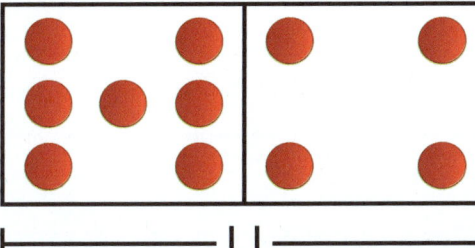

$\underline{11} \ominus \underline{7} = \underline{4}$

$\underline{7}$ pebbles

1. You have 12 erasers. Your friend takes some of them. You have 5 left. How many erasers does your friend take?

Circle what you know.   Underline what you need to find.

Solve:

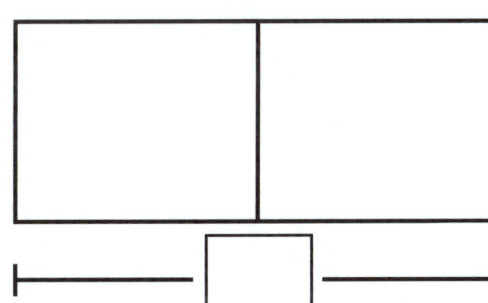

___ ◯ ___ = ___

___ erasers

2. You have 17 glitter pens. Your friend has 9 fewer than you. How many does your friend have?

___ glitter pens

Chapter 5 | Lesson 7   two hundred eighty-three   283

3. **DIG DEEPER!** A group of students are at a library. 6 of them leave. There are 7 left. Which part-part-whole model shows how many students were at the library to start?

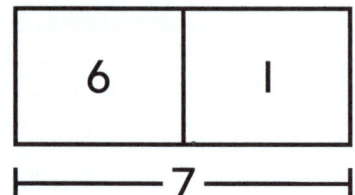 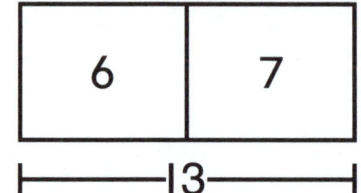

4. **Modeling Real Life** Your friend's shirt has 14 buttons. Your shirt has 7 fewer. How many buttons does your shirt have?

Friend:

You:

_____ buttons

Review & Refresh

5. There are 8 🦦.
   3 🦦 swim away.
   How many 🦦 are left?

   ___ − ___ = ___

   ___ 🦦

284   two hundred eighty-four

Name _____

Performance Task 5

1. You keep track of the number of honeybees and bumblebees you see.

Day	Honeybees
Monday	12
Tuesday	6
Wednesday	13

Day	Bumblebees
Monday	5
Tuesday	14
Wednesday	

a. How many more honeybees did you see on Monday than on Tuesday?

____ more honeybees

b. How many fewer bees did you see on Monday than on Tuesday?

____ fewer bees

c. How many bumblebees must you see on Wednesday so that the numbers of bees you see on Tuesday and Wednesday are the same?

____ bumblebees

Three in a Row: Subtraction

To Play: Place the Three in a Row: Subtraction Game Cards in a pile. Players take turns. On your turn, flip over the top card and solve the problem. Place a counter on the answer. Your turn is over. Repeat this process until a player gets three in a row.

Name _____

Chapter Practice 5

5.1 Count Back to Subtract within 20

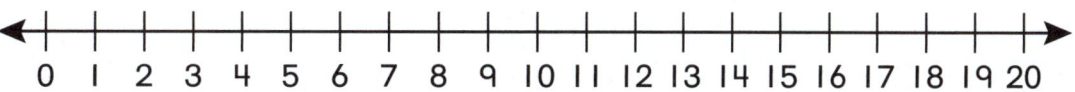

1. 11 − 3 = ___

2. 13 − 4 = ___

3. ___ = 15 − 3

4. ___ = 16 − 8

5.2 Use Addition to Subtract within 20

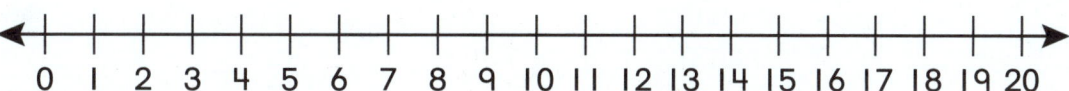

5. 11 − 9 = ?

   Think 9 + ___ = 11.

   So, 11 − 9 = ___.

6. 13 − 8 = ?

   Think 8 + ___ = 13.

   So, 13 − 8 = ___.

7. ___ = 20 − 7

8. ___ = 12 − 3

5.3 Subtract 9

Get to 10 to subtract.

9. 15 − 9 = ?

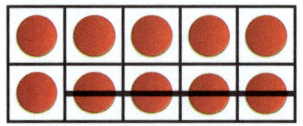

15 − ___ = 10

10 − ___ = ___

So, 15 − 9 = ___.

10. 17 − 9 = ?

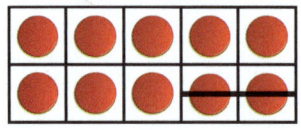

17 − ___ = 10

10 − ___ = ___

So, 17 − 9 = ___.

5.4 Get to 10 to Subtract

Get to 10 to subtract.

11. 12 − 7 = ?

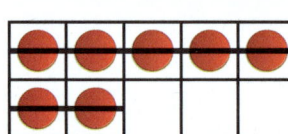

12 − ___ = 10

10 − ___ = ___

So, 12 − 7 = ___.

12. 17 − 8 = ?

17 − ___ = 10

10 − ___ = ___

So, 17 − 8 = ___.

5.5 More True or False Equations

Is the equation true or false?

13. $8 + 3 + 8 \stackrel{?}{=} 13 + 6$ $8 + 3 + 8:$ ___ $13 + 6:$ ___

___ $\stackrel{?}{=}$ ___

True   False

14. $16 - 8 \stackrel{?}{=} 6 + 2$

___ $\stackrel{?}{=}$ ___

True   False

15. $14 - 7 \stackrel{?}{=} 12 - 9$

___ $\stackrel{?}{=}$ ___

True   False

5.6 Make True Equations

16. $11 - 7 = 10 - ?$

___ $= 10 - ?$

___ $= 10 -$ ___

So, $11 - 7 = 10 -$ ___.

17. $2 + 0 + 6 = ? - 5$

___ $= ? - 5$

___ $=$ ___ $- 5$

So, $2 + 0 + 6 =$ ___ $- 5$.

5.7 Problem Solving: Subtraction within 20

18. There are 13 people on a train. Some of them exit. There are 5 left. How many people exit the train?

Circle what you know.

Underline what you need to find.

Solve:

___ ◯ ___ = ___   ___ people

19. A group of students are at a museum. 8 of them leave. There are 7 left. How many students were there to start?

___ students

20. You and a friend play basketball. Your friend scores 17 points. You score 8 fewer. How many points do you score?

___ points

6 Count and Write Numbers to 120

- What are some ways people raise money?
- How many quarters are there in all?

Chapter Learning Target:
Understand counting.

Chapter Success Criteria:
- I can identify numbers on a chart.
- I can describe numbers on a chart.
- I can count on from a number.
- I can write numbers.

Name ______________________

6 Vocabulary

Review Words
- column
- decade numbers
- hundred chart
- row

Organize It

Use the review words to complete the graphic organizer.

1	2	3	4	5	6	7	8	9	10
11	12	13	14	15	16	17	18	19	20
21	22	23	24	25	26	27	28	29	30
31	32	33	34	35	36	37	38	39	40
41	42	43	44	45	46	47	48	49	50
51	52	53	54	55	56	57	58	59	60
61	62	63	64	65	66	67	68	69	70
71	72	73	74	75	76	77	78	79	80
81	82	83	84	85	86	87	88	89	90
91	92	93	94	95	96	97	98	99	100

Define It

Use your vocabulary cards to identify the words.

23

Chapter 6 Vocabulary Cards

120 chart	column
decade numbers	digit
ones	ones place
row	tens

The digits of 16 are 1 and 6.

16

23

23 has 2 tens.

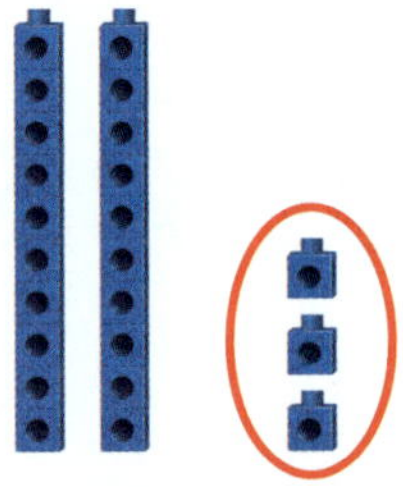

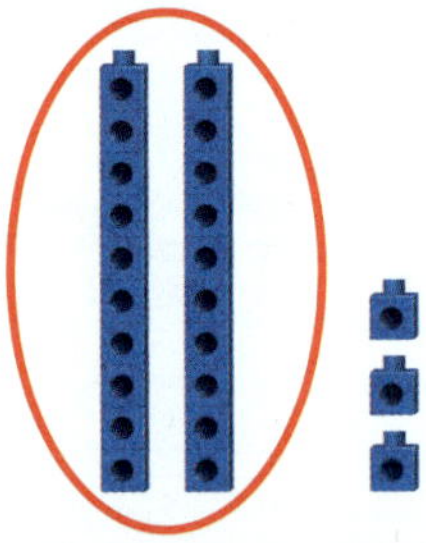

23 has 3 ones.

Chapter 6 Vocabulary Cards

tens place

23

Name _____

Learning Target: Count to 120 by ones.

Count to 120 by Ones

6.1

 Explore and Grow

Point to each number as you count to 120. Color the first two rows and the last two rows. How are the rows the same? How are they different?

1	2	3	4	5	6	7	8	9	10
11	12	13	14	15	16	17	18	19	20
21	22	23	24	25	26	27	28	29	30
31	32	33	34	35	36	37	38	39	40
41	42	43	44	45	46	47	48	49	50
51	52	53	54	55	56	57	58	59	60
61	62	63	64	65	66	67	68	69	70
71	72	73	74	75	76	77	78	79	80
81	82	83	84	85	86	87	88	89	90
91	92	93	94	95	96	97	98	99	100
101	102	103	104	105	106	107	108	109	110
111	112	113	114	115	116	117	118	119	120

Chapter 6 | Lesson 1

Think and Grow

Count by ones: fourteen, fifteen, sixteen, seventeen, eighteen, nineteen

14, 15, 16, 17, 18, 19

1	2	3	4	5	6	7	8	9	10
11	12	13	14	15	16	17	18	19	20
21	22	23	24	25	26	27	28	29	30
31	32	33	34	35	36	37	38	39	40
41	42	43	44	45	46	47	48	49	50
51	52	53	54	55	56	57	58	59	60	← row
61	62	63	64	65	66	67	68	69	70
71	72	73	74	75	76	77	78	79	80
81	82	83	84	85	86	87	88	89	90
91	92	93	94	95	96	97	98	99	100
101	102	103	104	105	106	107	108	109	110
111	112	113	114	115	116	117	118	119	120

120 chart

Show and Grow   *I can do it!*

Count by ones to write the missing numbers.

1. 82, ____, ____, ____, ____, ____

2. 103, ____, ____, ____, ____, ____

Name _____

**Apply and Grow: Practice**

Count by ones to write the missing numbers.

3. 56, ____, ____, ____, ____, ____

4. 98, ____, ____, ____, ____, ____

5. 115, ____, ____, ____, ____, ____

6. ____, ____, 42, ____, ____, ____

Write the missing numbers in the chart.

7.
32		34	
	43		45

8.
	82		
91		93	

9. **Structure** Write a number between 95 and 105. Then count by ones to write the next 7 numbers.

____, ____, ____, ____, ____, ____, ____, ____

Chapter 6 | Lesson 1   two hundred ninety-five 295

Think and Grow: Modeling Real Life

You have 108 bouncy balls. You want 112. How many more bouncy balls do you need?

Draw more balls to show 112:

_____ more bouncy balls

Show and Grow — I can think deeper!

10. You have 66 rocks. You want 75. How many more rocks do you need?

    Draw more rocks to show 75:

    _____ more rocks

Name _____

Practice 6.1

Learning Target: Count to 120 by ones.

1	2	3	4	5	6	7	8	9	10
11	12	13	14	15	16	17	18	19	20
21	22	23	24	25	26	27	28	29	30
31	32	33	34	35	36	37	38	39	40
41	42	43	44	45	46	47	48	49	50
51	52	53	54	55	56	57	58	59	60
61	62	63	64	65	66	67	68	69	70
71	72	73	74	75	76	77	78	79	80
81	82	83	84	85	86	87	88	89	90
91	92	93	94	95	96	97	98	99	100
101	102	103	104	105	106	107	108	109	110
111	112	113	114	115	116	117	118	119	120

72, **73**, **74**, **75**, **76**, **77**

Count by ones to write the missing numbers.

1. 57, ____, ____, ____, ____, ____

2. 109, ____, ____, ____, ____, ____

3. 40, ____, ____, ____, ____, ____

4. ____, ____, ____, 100, ____, ____

Chapter 6 | Lesson 1   two hundred ninety-seven   297

Write the missing numbers in the chart.

5.

20		22	
	31		33

6.

	103		
112		114	

7. **Structure** Write a number between 85 and 95. Then count by ones to write the next 7 numbers.

_____ , _____ , _____ , _____ , _____ , _____ , _____ , _____

8. **Modeling Real Life** There are 110 tokens. You want 119. How many more tokens do you need?

_____ more tokens

Review & Refresh

9. ? − 6 = 4

| 6 | 4 |

Think 6 + 4 = _____.

So, _____ − 6 = 4.

298   two hundred ninety-eight

Name _____

Count to 120 by Tens 6.2

Learning Target: Count to 120 by tens.

 **Explore and Grow**

Count to 10. Circle the number. Count 10 more. Circle the number. Continue until you reach 120.

1	2	3	4	5	6	7	8	9	10
11	12	13	14	15	16	17	18	19	20
21	22	23	24	25	26	27	28	29	30
31	32	33	34	35	36	37	38	39	40
41	42	43	44	45	46	47	48	49	50
51	52	53	54	55	56	57	58	59	60
61	62	63	64	65	66	67	68	69	70
71	72	73	74	75	76	77	78	79	80
81	82	83	84	85	86	87	88	89	90
91	92	93	94	95	96	97	98	99	100
101	102	103	104	105	106	107	108	109	110
111	112	113	114	115	116	117	118	119	120

Chapter 6 | Lesson 2

two hundred ninety-nine

Think and Grow

40, 50, 60, 70, 80, 90

1	2	3	4	5	6	7	8	9	10
11	12	13	14	15	16	17	18	19	20
21	22	23	24	25	26	27	28	29	30
31	32	33	34	35	36	37	38	39	40
41	42	43	44	45	46	47	48	49	50
51	52	53	54	55	56	57	58	59	60
61	62	63	64	65	66	67	68	69	70
71	72	73	74	75	76	77	78	79	80
81	82	83	84	85	86	87	88	89	90
91	92	93	94	95	96	97	98	99	100
101	102	103	104	105	106	107	108	109	110
111	112	113	114	115	116	117	118	119	120

↑ column

↑ decade numbers

Count by tens: forty, fifty, sixty, seventy, eighty, ninety

Show and Grow   I can do it!

Count by tens to write the missing numbers.

1. 70, _____, _____, _____, _____, _____

2. 31, _____, _____, _____, _____, _____

Name _____

 **Apply and Grow: Practice**

Count by tens to write the missing numbers.

3. 62, _____, _____, _____, _____, _____

4. 43, _____, _____, _____, _____, _____

5. _____, _____, 30, _____, _____, _____

6. Write the missing numbers from the chart. Then count on by tens to write the next three numbers.

?	72	73	74	75	76	77	78	79	80
?	82	83	84	85	86	87	88	89	90

_____, _____, _____, _____, _____

7. **YOU BE THE TEACHER** Your friend counts by tens starting with 27. Is your friend correct? Show how you know.

27, 37, 47, 67, 77, 87

Chapter 6 | Lesson 2   three hundred one   **301**

Think and Grow: Modeling Real Life

You have 50 points. On your next turn, you knock over 6 cans. How many points do you have now?

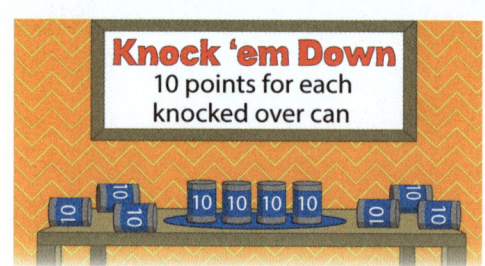

Write the numbers:

_____ points

Show and Grow   I can think deeper!

8. You have 21 points. On your next turn, 3 beanbags land in the circle. How many points do you have now?

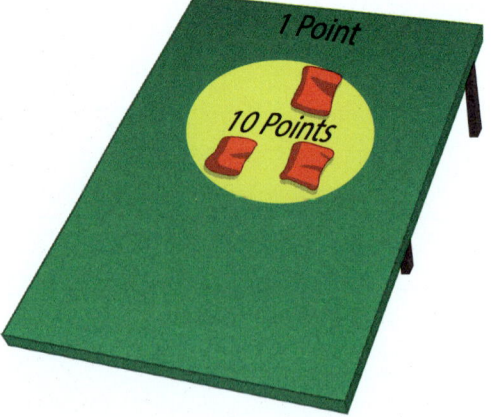

Write the numbers:

_____ points

Name _____

Practice 6.2

Learning Target: Count to 120 by tens.

[Hundreds chart 1–120]

45, __55__, __65__, __75__, __85__, __95__

Count by tens to write the missing numbers.

1. 69, ____, ____, ____, ____, ____

2. 41, ____, ____, ____, ____, ____

3. 16, ____, ____, ____, ____, ____

4. ____, ____, ____, 94, ____, ____

5. Write the missing numbers from the chart. Then count on by tens to write the next three numbers.

1	2	3	4	5		7	8	9	10
11	12	13	14	15		17	18	19	20

____, ____, ____, ____, ____

6. **DIG DEEPER!** You count to 50. You only count 5 numbers. Did you count by ones or by tens? Show how you know.

7. **Modeling Real Life** You have 30 points. On your next turn, 4 balls stick to the wall. How many points do you have now?

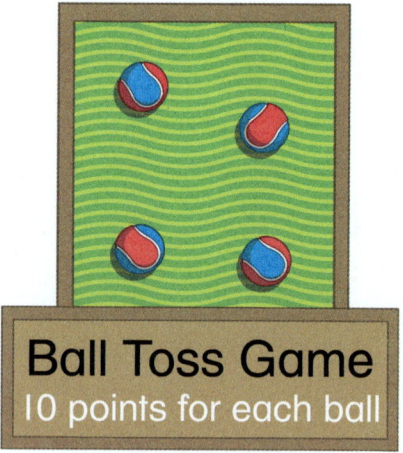

Ball Toss Game
10 points for each ball

_____ points

Review & Refresh

8. 3 + 1 = _____   **9.** 5 − 1 = _____

Name _____

Learning Target: Understand and write numbers from 11 to 19.

Compose Numbers 11 to 19

6.3

Explore and Grow

Color to show 13 and 17. What is the same about the numbers? What is different?

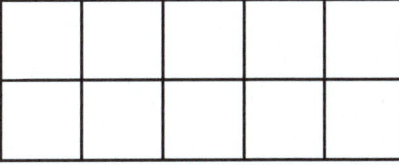

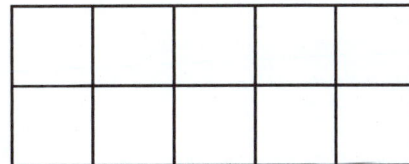

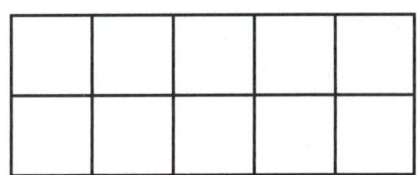

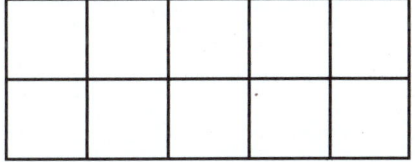

Chapter 6 | Lesson 3

Think and Grow

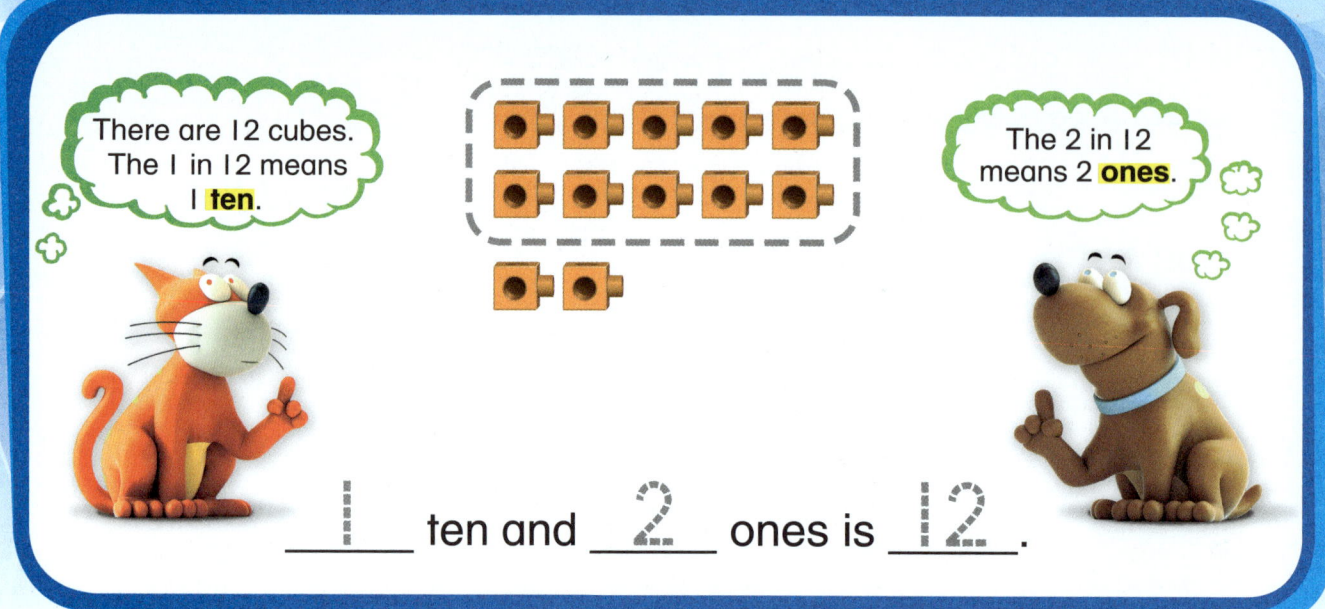

___1___ ten and ___2___ ones is ___12___.

Show and Grow — I can do it!

1. Circle 10 feathers. Complete the sentence.

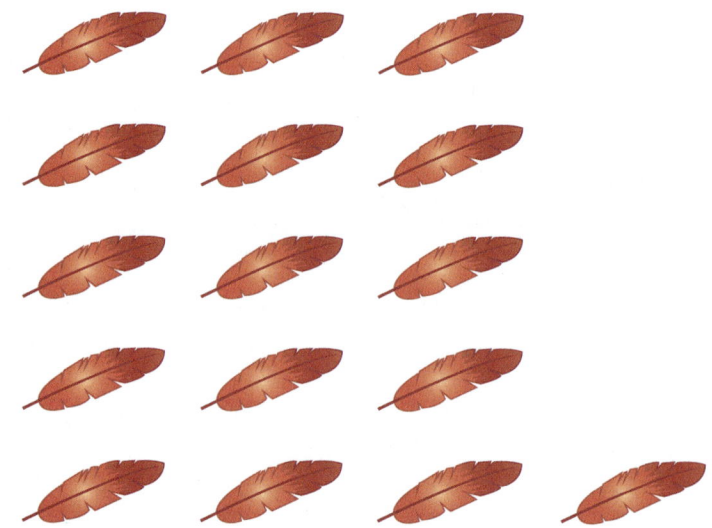

_____ ten and _____ ones is _____.

Name _____

✓ Apply and Grow: Practice

Circle 10 objects. Complete the sentence.

2.

_____ ten and _____ ones is _____.

3.

_____ ten and _____ ones is _____.

4.

_____ ten and _____ ones is _____.

5. **Number Sense** Color to show the number. Complete the sentence.

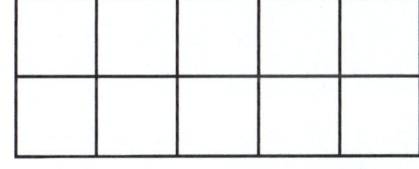 

_____ ten and _____ ones is _____.

Chapter 6 | Lesson 3   three hundred seven   307

# Think and Grow: Modeling Real Life

You have 15 footballs. A bag can hold 10. You fill a bag. How many footballs are *not* in the bag?

Draw a picture:

Write the missing numbers: _____ ten and _____ ones

_____ footballs

Show and Grow I can think deeper!

6. Your teacher has 18 calculators. A case can hold 10. Your teacher fills a case. How many calculators are *not* in the case?

Draw a picture:

Write the missing numbers: _____ ten and _____ ones

_____ calculators

Name _____

Practice 6.3

Learning Target: Understand and write numbers from 11 to 19.

__1__ ten and __7__ ones is __17__.

Circle 10 objects. Complete the sentence.

1. _____ ten and _____ one is _____.

2.

_____ ten and _____ ones is _____.

3.

_____ ten and _____ ones is _____.

Chapter 6 | Lesson 3

4. **Number Sense** Color to show the number. Complete the sentence.

_____ ten and _____ ones is _____.

5. **Number Sense** Match.

1 ten and 3 ones	1 ten and 8 ones	12 ones
13	12	18

6. **Modeling Real Life** You have 16 books. A backpack can hold 10. You fill a backpack. How many books are *not* in the backpack?

_____ books

Review & Refresh

7. 10 + 0 = _____

8. 10 + 10 = _____

310 three hundred ten

Name _____

Learning Target: Understand and write decade numbers.

Explore and Grow

Circle groups of 10. Write the number of groups.

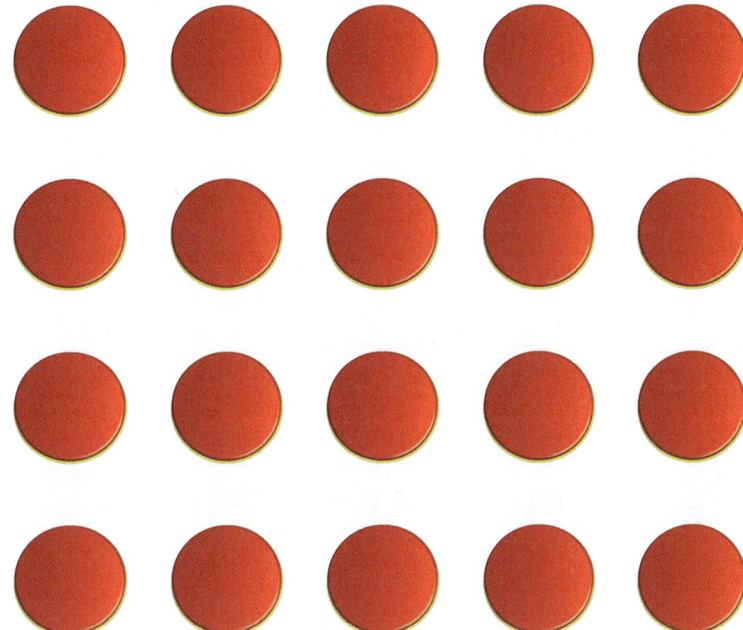

_____ groups

How many counters are there in all?

_____ counters

Chapter 6 | Lesson 4

 **Think and Grow**

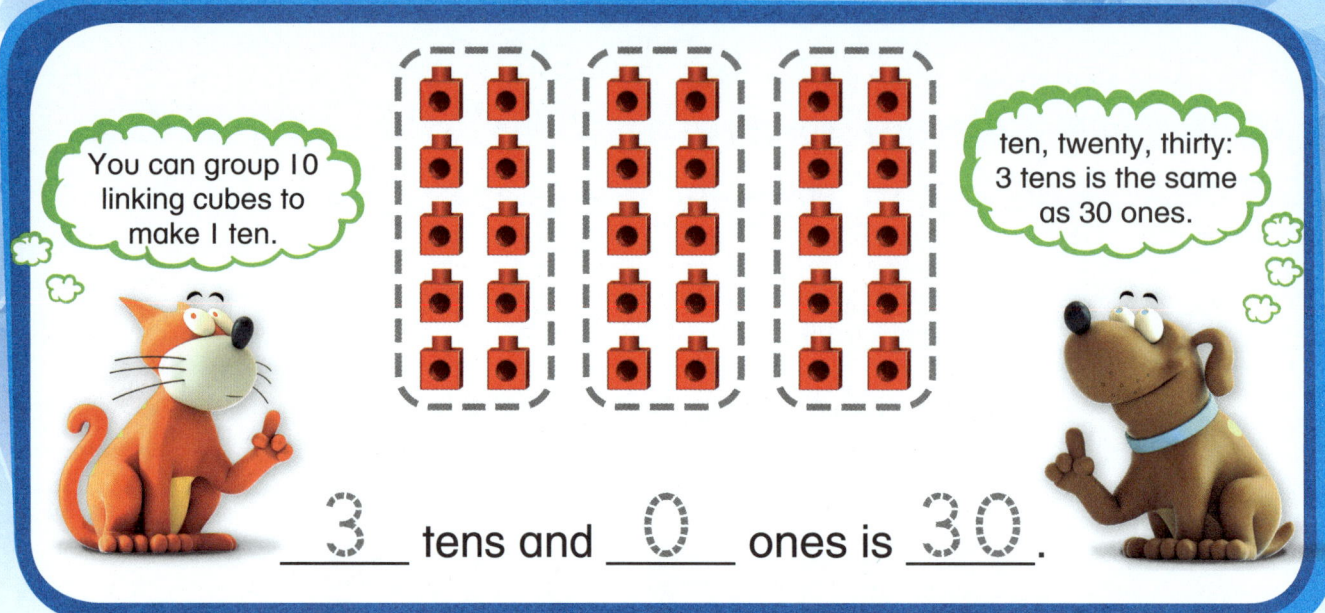

___3___ tens and ___0___ ones is ___30___.

Show and Grow I can do it!

Circle groups of 10. Complete the sentence.

1.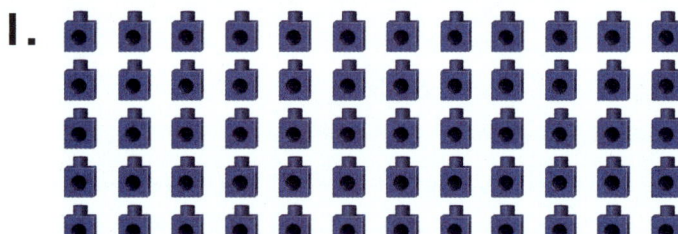

   _____ tens and _____ ones is _____.

2.

   _____ tens and _____ ones is _____.

Name _____

✓ Apply and Grow: Practice

Circle groups of 10. Complete the sentence.

3.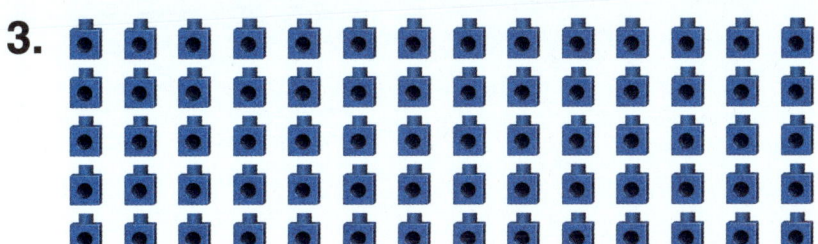

_____ tens and _____ ones is _____.

4.

_____ tens and _____ ones is _____.

5.

_____ ten and _____ ones is _____.

6. **Number Sense** You have 4 groups of 10 linking cubes. How many linking cubes do you have?

_____ linking cubes

Chapter 6 | Lesson 4

Think and Grow: Modeling Real Life

You read 10 books every month. You want to read 40 books. How many months does it take?

Draw a picture:

Write the missing numbers: _____ tens and _____ ones

_____ months

Show and Grow *I can think deeper!*

7. There are 10 dog bones in each box. You need 20 bones. How many boxes do you need?

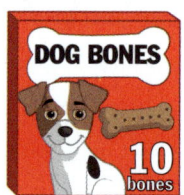

Draw a picture:

Write the missing numbers: _____ tens and _____ ones

_____ boxes

Name _____

Practice 6.4

Learning Target: Understand and write decade numbers.

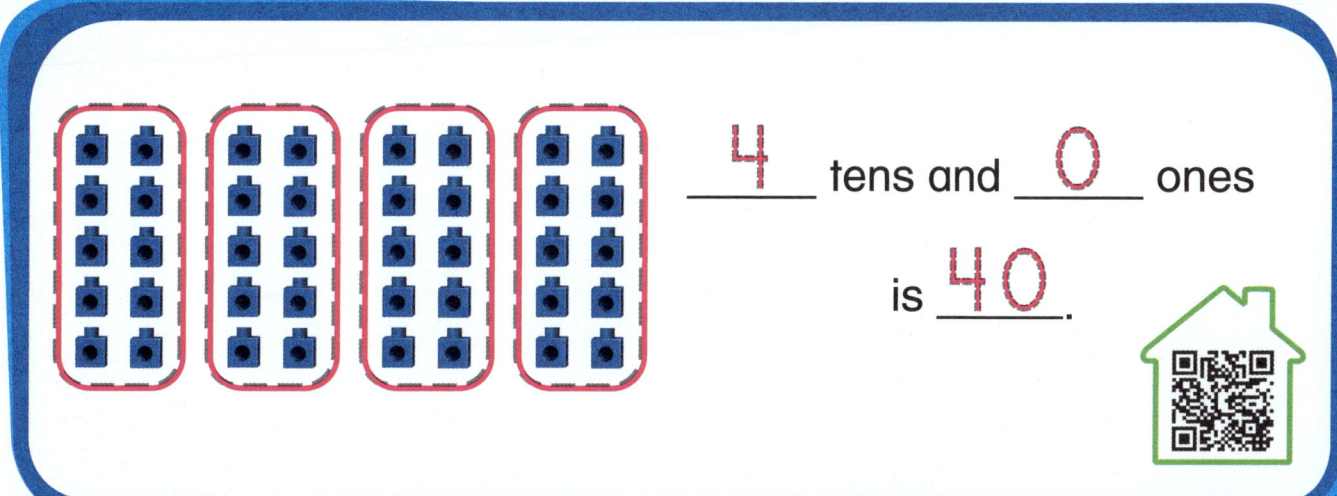

__4__ tens and __0__ ones is __40__.

Circle groups of 10. Complete the sentence.

1.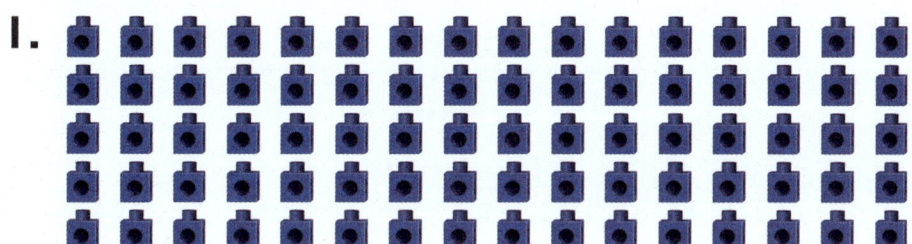

   _____ tens and _____ ones is _____.

2. _____ tens and _____ ones is _____.

3. _____ tens and _____ ones is _____.

Chapter 6 | Lesson 4   three hundred fifteen 315

Circle groups of 10. Complete the sentence.

4.

_____ tens and _____ ones is _____.

5. **Number Sense** You have 7 groups of 10 linking cubes. How many linking cubes do you have?

_____ linking cubes

6. **Modeling Real Life** You swim 10 laps at every practice. You want to swim 50 laps. How many practices will it take?

_____ practices

Review & Refresh

7. $4 + 3 + 4 =$ _____

8. $1 + 5 + 9 =$ _____

9. $2 + 2 + 1 =$ _____

10. $7 + 3 + 6 =$ _____

Name _____

Learning Target: Count tens and ones to write numbers.

Tens and Ones 6.5

 Explore and Grow

Model 2 tens and 3 ones. Write the number.

Tens	Ones

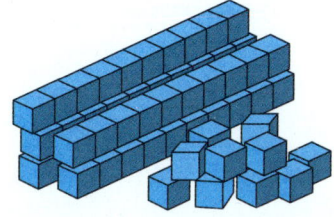

_____

Chapter 6 | Lesson 5

three hundred seventeen 317

 **Think and Grow**

Tens	Ones	→	Tens	Ones
			4	5

> The 4 in 45 is in the **tens place**. The 5 in 45 is in the **ones place**.

__4__ tens and __5__ ones is __45__.

Show and Grow   I can do it!

1.

Tens	Ones	→	Tens	Ones

_____ tens and _____ one is _____.

318   three hundred eighteen

Name _____

✓ Apply and Grow: Practice

2.

Tens	Ones

_____ tens and _____ ones is _____.

3.

Tens	Ones

_____ tens and _____ ones is _____.

4.

Tens	Ones

_____ tens and _____ ones is _____.

5. **YOU BE THE TEACHER** You have 92 linking cubes. Your friend says that there are 2 tens and 9 ones. Is your friend correct? Show how you know.

Chapter 6 | Lesson 5

three hundred nineteen 319

# Think and Grow: Modeling Real Life

Your teacher has 2 packages of dice and 3 extra dice. Each package has 10 dice. How many dice are there in all?

Draw a picture:

Write the missing numbers: _____ tens and _____ ones

_____ dice

Show and Grow   I can think deeper!

6. You have 3 boxes of colored pencils and 4 extra colored pencils. Each box has 10 pencils. How many colored pencils are there in all?

Draw a picture:

Write the missing numbers: _____ tens and _____ ones

_____ colored pencils

Name _____

Practice 6.5

Learning Target: Count tens and ones to write numbers.

Tens	Ones
(5 tens rods)	(3 ones cubes)

→

Tens	Ones
5	3

__5__ tens and __3__ ones is __53__.

1.

Tens	Ones
(6 tens rods)	(9 ones cubes)

→

Tens	Ones

_____ tens and _____ ones is _____.

2.

(6 tens rods and 1 one cube)

Tens	Ones

_____ tens and _____ one is _____.

Chapter 6 | Lesson 5   three hundred twenty-one 321

3.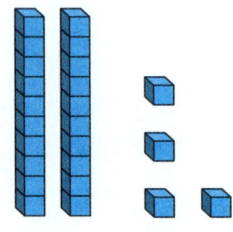

Tens	Ones

_____ tens and _____ ones is _____.

4. **YOU BE THE TEACHER** You have 17 linking cubes. Your friend says that there is 1 ten and 7 ones. Is your friend correct? Show how you know.

5. **Modeling Real Life** You have 5 bags of apples and 1 extra apple. Each bag has 10 apples. How many apples are there in all?

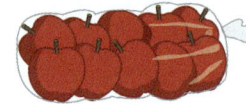

_____ apples

Review & Refresh

6. _____ + 6 = 10

7. _____ + 2 = 8

Name _____

Learning Target: Use quick sketches to model numbers as tens and ones.

Make Quick Sketches 6.6

Explore and Grow

Model the number 26.

Tens	Ones

Chapter 6 | Lesson 6

three hundred twenty-three 323

Think and Grow

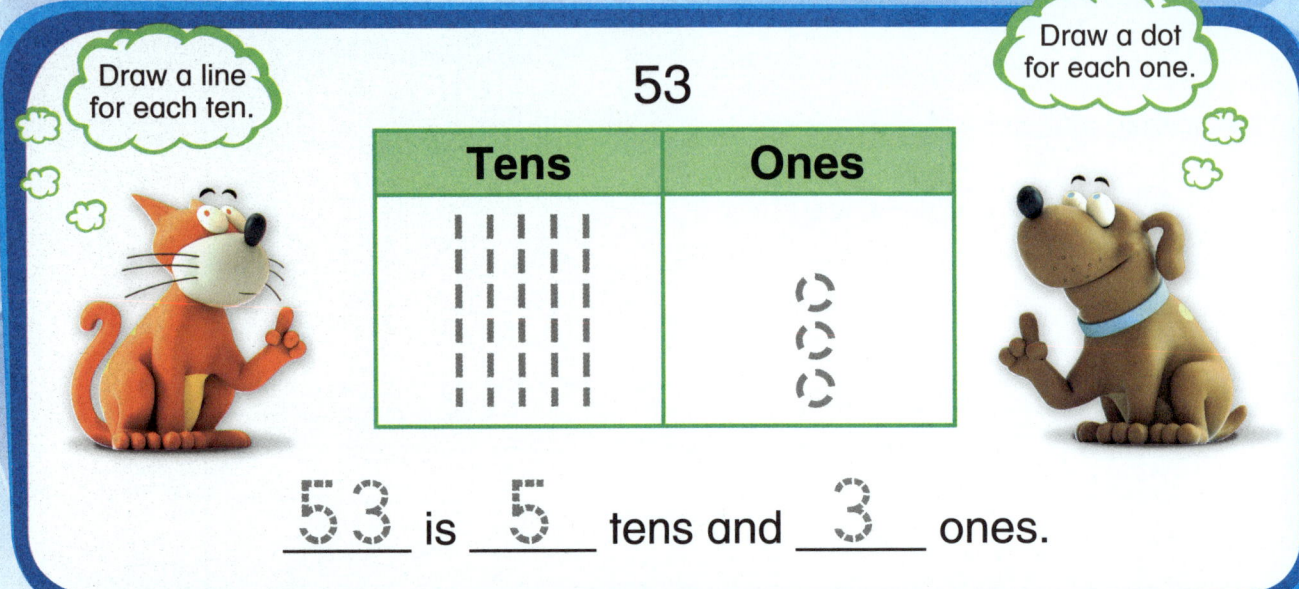

53 is 5 tens and 3 ones.

Show and Grow — I can do it!

Make a quick sketch. Complete the sentence.

1. 72

Tens	Ones

_____ is _____ tens and _____ ones.

2. 36

Tens	Ones

_____ is _____ tens and _____ ones.

Name _____

 Apply and Grow: Practice

Make a quick sketch. Complete the sentence.

3. 45

Tens	Ones

_____ is _____ tens and _____ ones.

4. 87

Tens	Ones

_____ is _____ tens and _____ ones.

5. 64

Tens	Ones

_____ is _____ tens and _____ ones.

6. **DIG DEEPER!** Which sketch shows 54?

Chapter 6 | Lesson 6   three hundred twenty-five   **325**

 **Think and Grow: Modeling Real Life**

You need 58 plates for a party. You have 51. How many more plates do you need?

Complete the model:

_____ more plates

Show and Grow I can think deeper!

7. You need 80 tickets for a prize. You have 73. How many more tickets do you need?

Complete the model:

_____ more tickets

Name _____

Practice 6.6

Learning Target: Use quick sketches to model numbers as tens and ones.

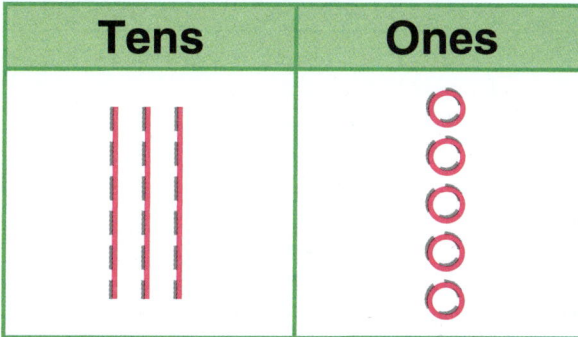

35

35 is 3 tens and 5 ones.

Make a quick sketch. Complete the sentence.

1. 27

Tens	Ones

_____ is _____ tens and _____ ones.

2. 61

Tens	Ones

_____ is _____ tens and _____ one.

Chapter 6 | Lesson 6   three hundred twenty-seven   327

Make a quick sketch. Complete the sentence.

3. 92

Tens	Ones

_____ is _____ tens and _____ ones.

4. **DIG DEEPER!** Which sketch shows 87?

5. **Modeling Real Life** You need 55 beads to make a necklace. You have 48. How many more beads do you need?

_____ more beads

Review & Refresh

6. Color the shapes that have only 4 sides.

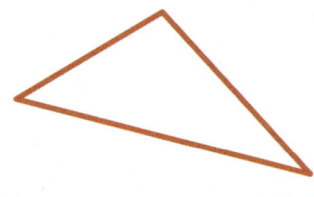 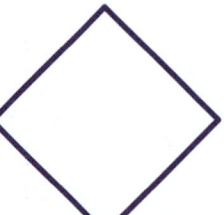 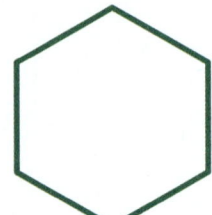 

Name _____

Learning Target: Understand the value of each digit in a two-digit number.

Understand Place Value 6.7

Explore and Grow

Newton has 2 rods. Make a quick sketch. Write the number.

_____

Descartes has 2 cubes. Make a quick sketch. Write the number.

_____

How are the models alike? How are they different?

Chapter 6 | **Lesson 7**

Think and Grow

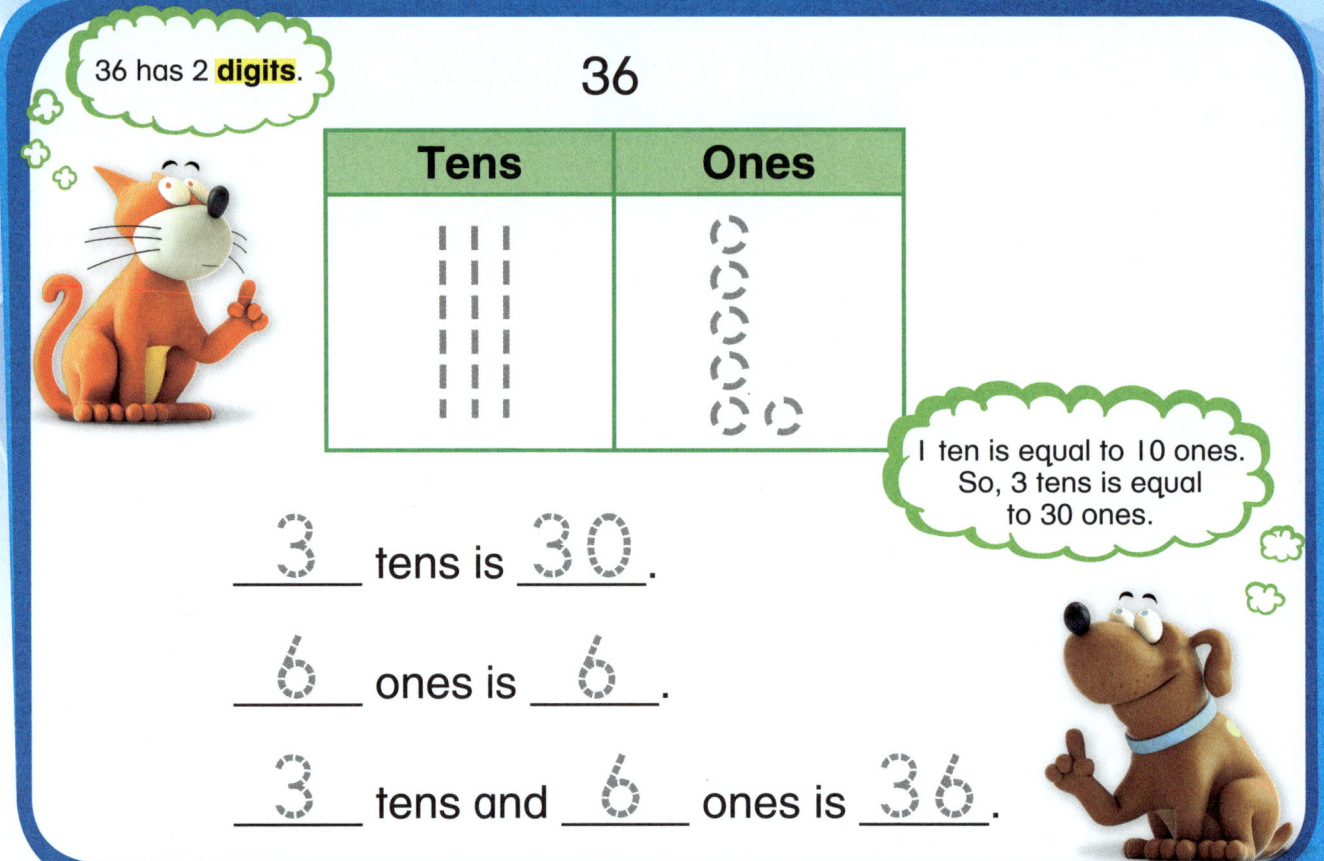

36 has 2 **digits**.

36

I ten is equal to 10 ones. So, 3 tens is equal to 30 ones.

___3___ tens is ___30___.

___6___ ones is ___6___.

___3___ tens and ___6___ ones is ___36___.

Show and Grow   I can do it!

1. Make a quick sketch. Complete the sentences.

   64

Tens	Ones

   _____ tens is _____.

   _____ ones is _____.

   _____ tens and _____ ones is _____.

Name _____

 Apply and Grow: Practice

Make a quick sketch. Complete the sentences.

2. 72

Tens	Ones

_____ tens is _____.

_____ ones is _____.

_____ tens and _____ ones is _____.

3. 98

Tens	Ones

_____ tens is _____.

_____ ones is _____.

_____ tens and _____ ones is _____.

4. 57

_____ tens is _____.

_____ ones is _____.

_____ tens and _____ ones is _____.

Chapter 6 | Lesson 7

Think and Grow: Modeling Real Life

You have 94 charms to make bracelets. There are 10 charms on each bracelet. How many bracelets can you make?

Model:

Write the missing numbers: _____ tens and _____ ones

_____ bracelets

Show and Grow   I can think deeper!

5. You have 67 seeds. You plant 10 seeds in a row. How many rows can you plant?

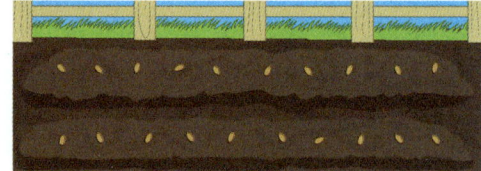

Model:

Write the missing numbers: _____ tens and _____ ones

_____ rows

Name _____

Practice 6.7

Learning Target: Understand the value of each digit in a two-digit number.

25

Tens	Ones
‖	ooooo

__2__ tens is __20__.

__5__ ones is __5__.

__2__ tens and __5__ ones is __25__.

Make a quick sketch. Complete the sentences.

1. 81

Tens	Ones

_____ tens is _____.

_____ one is _____.

_____ tens and _____ one is _____.

2. 53

Tens	Ones

_____ tens is _____.

_____ ones is _____.

_____ tens and _____ ones is _____.

Chapter 6 | Lesson 7

3. 49

   _____ tens is _____.

   _____ ones is _____.

   _____ tens and _____ ones is _____.

4. 76

   _____ tens is _____.

   _____ ones is _____.

   _____ tens and _____ ones is _____.

5. **Modeling Real Life** You have 77 crayons. A box can hold 10 crayons. How many boxes can you fill?

_____ boxes

Review & Refresh

Is the equation true or false?

6. $4 + 9 \stackrel{?}{=} 2 + 3 + 5$

   True   False

7. $5 + 3 \stackrel{?}{=} 4 + 4$

   True   False

Name _____

Learning Target: Show different ways to write numbers.

Write Numbers in Different Ways 6.8

Explore and Grow

Model 27 two ways.

Tens	Ones

_____ tens and _____ ones is _____.

_____ tens and _____ ones is _____.

Chapter 6 | Lesson 8

three hundred thirty-five 335

Think and Grow

Model 46 two ways.

Tens	Ones
IIII	ooooo ooooo

__4__ tens and __6__ ones is 46.

Tens	Ones
III	ooooo ooooo ooooo o

Remember, 1 ten is the same as 10 ones.

__3__ tens and __16__ ones is 46.

Show and Grow — I can do it!

1. Model 25 two ways.

Tens	Ones

_____ tens and _____ ones is 25.

Tens	Ones

_____ tens and _____ ones is 25.

336 three hundred thirty-six

Name _____

Apply and Grow: Practice

2. Model 52 two ways.

Tens	Ones

_____ tens and _____ ones is 52.

Tens	Ones

_____ tens and _____ ones is 52.

3. Model 14 two ways.

Tens	Ones

_____ ten and _____ ones is 14.

Tens	Ones

_____ ten and _____ ones is 14.

4. **DIG DEEPER!** Circle all of the ways that show 39.

3 + 9   2 tens and 19 ones   9 tens and 3 ones

10 + 29   3 tens and 19 ones   39 ones

Chapter 6 | Lesson 8   three hundred thirty-seven 337

Think and Grow: Modeling Real Life

The models show how many seashells you and your friend have. Does your friend have the same number of seashells as you?

You   Friend

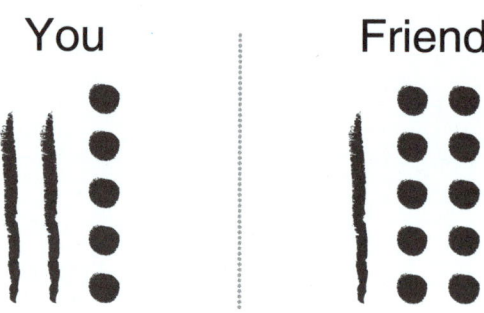

Circle: Yes   No

Show how you know:

Show and Grow — I can think deeper!

5. The models show how many erasers you and your friend have. Does your friend have the same number of erasers as you?

You   Friend

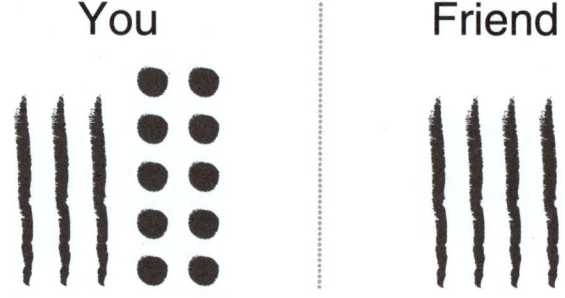

Circle: Yes   No

Show how you know:

Name _____

Practice 6.8

Learning Target: Show different ways to write numbers.

Model 18 two ways.

Tens	Ones
1	○○○○○○○○○ (8)

__1__ ten and
__8__ ones is 18.

Tens	Ones
	○○○○○○○○○○○○○○○○○○ (18)

__0__ tens and
__18__ ones is 18.

1. Model 49 two ways.

Tens	Ones

_____ tens and
_____ ones
is 49.

Tens	Ones

_____ tens and
_____ ones
is 49.

Chapter 6 | Lesson 8

2. **DIG DEEPER!** Circle all of the ways that show 45.

   40 + 5   45 tens and 0 ones   4 tens and 5 ones

   20 + 15   2 tens and 25 ones   54 ones

3. **Modeling Real Life** The models show the number of toy cars you and your friend have. Does your friend have the same number of toy cars as you?

   You    Friend

   Circle: Yes   No

   Show how you know:

   Review & Refresh

   Circle the heavier object.

   4.    5.

   340   three hundred forty

Name _____

Learning Target: Count and write numbers to 120.

Count and Write Numbers to 120

Explore and Grow

How many balls are there? How did you count?

_____ balls

Chapter 6 | Lesson 9

three hundred forty-one 341

Think and Grow

Count by tens. Then count by ones.

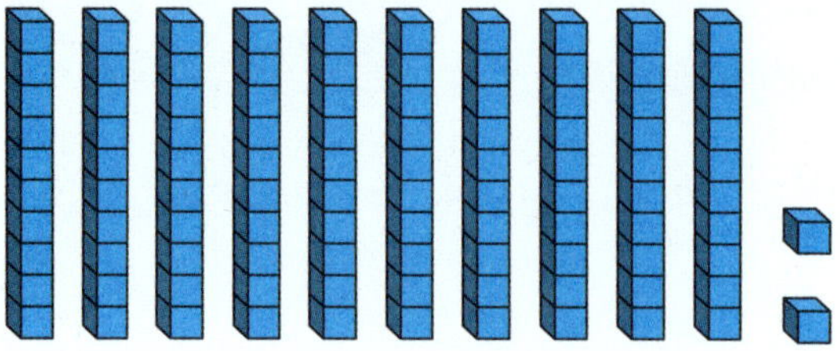

10 tens is equal to 100.

102

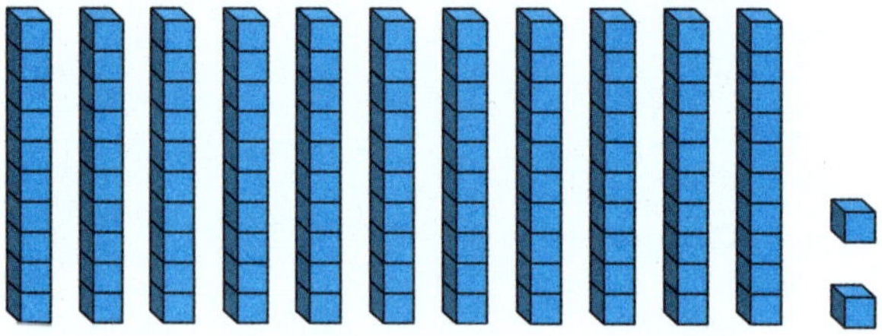

112

Show and Grow I can do it!

1.

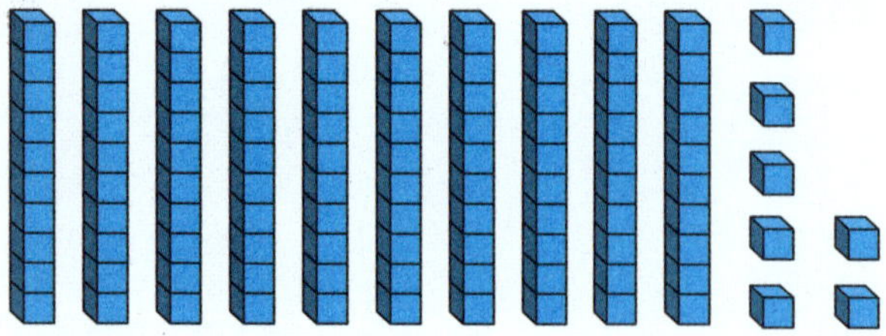

2.

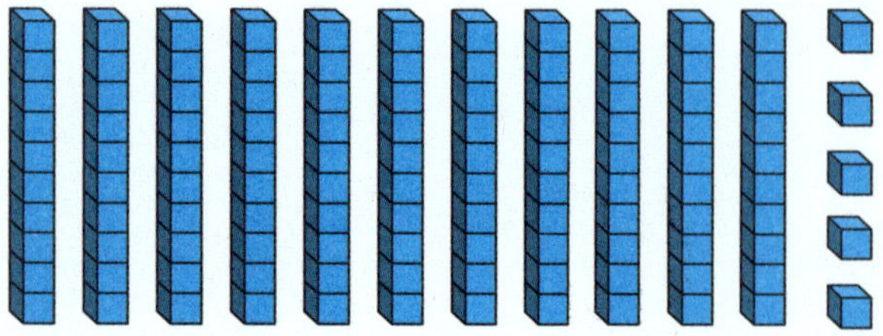

Apply and Grow: Practice

3.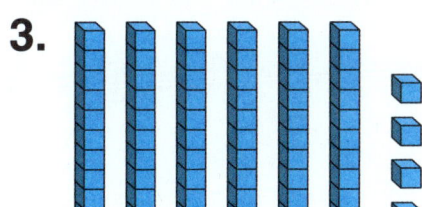

_____

4.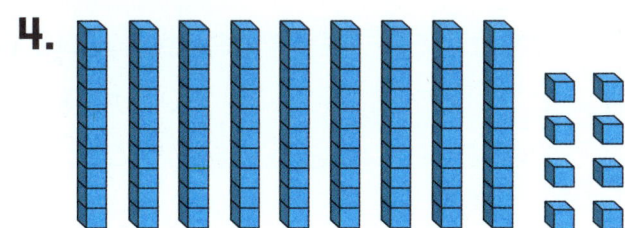

_____

5.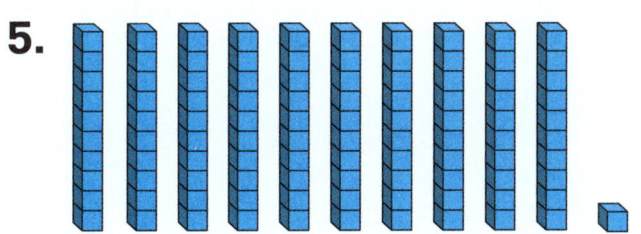

_____

6.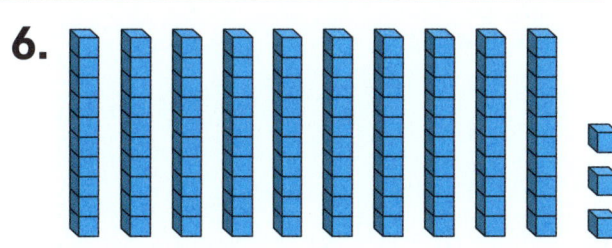

_____

7.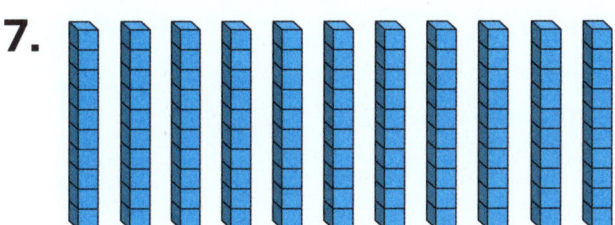

_____

8. **DIG DEEPER!** What number is equal to 10 tens and 8 ones? Show how you know.

# Think and Grow: Modeling Real Life

Your teacher has 12 bags of balloons. Each bag has 10 balloons. How many balloons are there in all?

Model:

_____ balloons

Show and Grow — I can think deeper!

9. A dentist has 10 boxes of toothbrushes and 9 extra toothbrushes. Each box has 10 toothbrushes. How many toothbrushes are there in all?

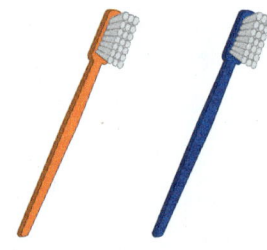

Model:

_____ toothbrushes

Name _____

Practice 6.9

Learning Target: Count and write numbers to 120.

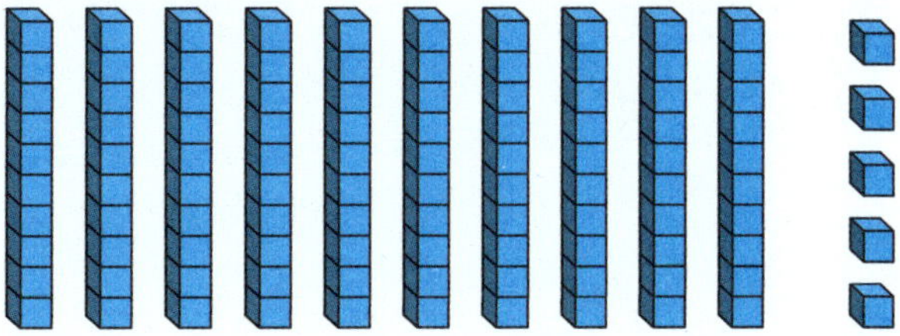

105

1. ____

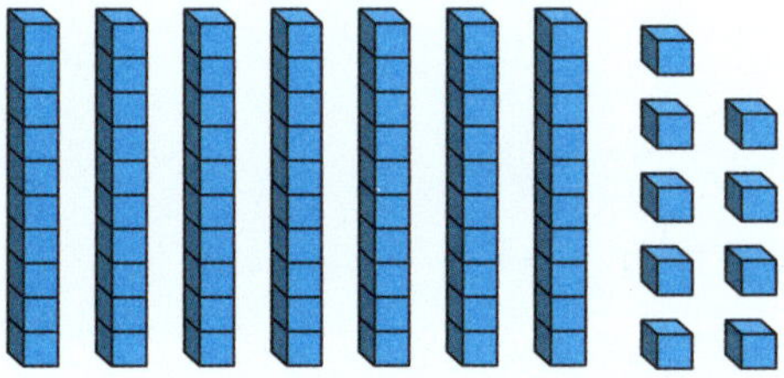

2. ____

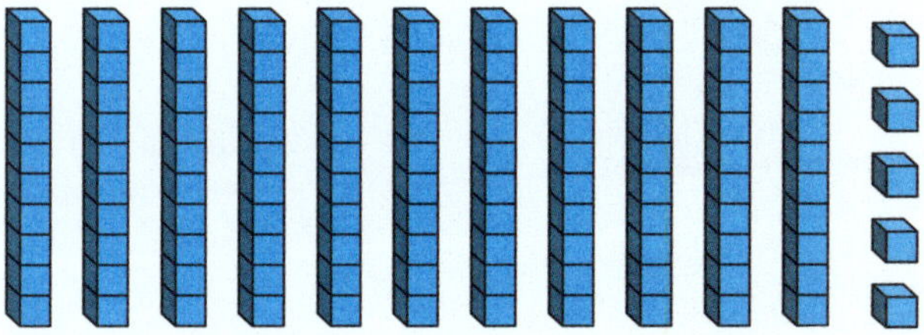

3. ____

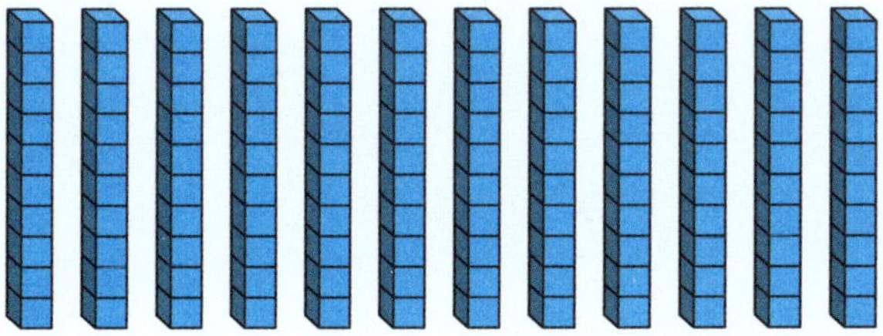

4. ____

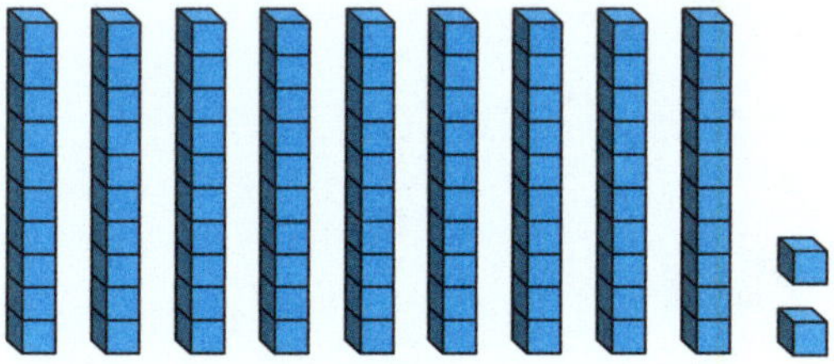

Chapter 6 | Lesson 9 — three hundred forty-five — 345

5. **DIG DEEPER!** What number is equal to 11 tens and 2 ones? Show how you know.

6. **Modeling Real Life** You have 4 packs of baseball cards and 6 packs of football cards. Each pack has 10 cards. How many cards do you have in all?

_____ cards

Review & Refresh

Make a 10 to add.

7. $6 + 5$

   $6 + \underline{} + \underline{}$

   $10 + \underline{} = \underline{}$

   So, $6 + 5 = \underline{}$.

8. $7 + 8$

   $7 + \underline{} + \underline{}$

   $10 + \underline{} = \underline{}$

   So, $7 + 8 = \underline{}$.

Name _____

Performance Task 6

1. Your class sells candles for a fundraiser. You earn 10 dollars for every large candle you sell and 1 dollar for every small candle.

   a. You sell 6 large candles. How much money do you raise?

   _____ dollars

   b. You want to raise 72 dollars. How much more money do you need to raise?

   _____ more dollars

   c. You also sell 12 small candles. Do you reach your goal?

   Yes   No

   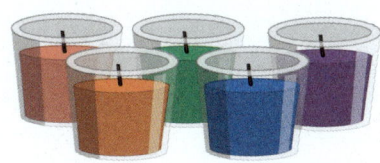

2. Your friend wants to raise 54 dollars. What are two ways your friend can sell large and small candles to reach her goal?

   _____ large candles and _____ small candles

   _____ large candles and _____ small candles

Chapter 6   three hundred forty-seven   347

Drop and Build

To Play: Take turns. On your turn, toss a cube onto a 120 chart so that your partner cannot see. Build the number you land on with base ten blocks. Have your partner say and write the number. Use the 120 chart to check the answer. Play until each partner builds three numbers.

Tens	Ones

Name _____

Chapter Practice 6

6.1 Count to 120 by Ones

1. Count by ones to write the missing numbers.

    99, _____, _____, _____, _____, _____

Write the missing numbers.

2.
68		70	
	79		81

3.
100		102	
	111		

6.2 Count to 120 by Tens

4. Write the missing numbers from the chart. Then count on by tens to write the next two numbers.

81	82	83	?	85	86	87	88	89	90
91	92	93	?	95	96	97	98	99	100

_____, _____, _____, _____

5. **YOU BE THE TEACHER** Your friend counts by tens starting with 53. Is your friend correct? Show how you know.

    53, 63, 73, 83, 103

6.3 Compose Numbers 11 to 19

6. Circle 10 ducks. Complete the sentence.

_____ ten and _____ ones is _____.

7. Modeling Real Life You have 19 tennis balls. A bag can hold 10. You fill a bag. How many tennis balls are *not* in the bag?

_____ tennis balls

6.4 Tens

8. Circle groups of 10. Complete the sentence.

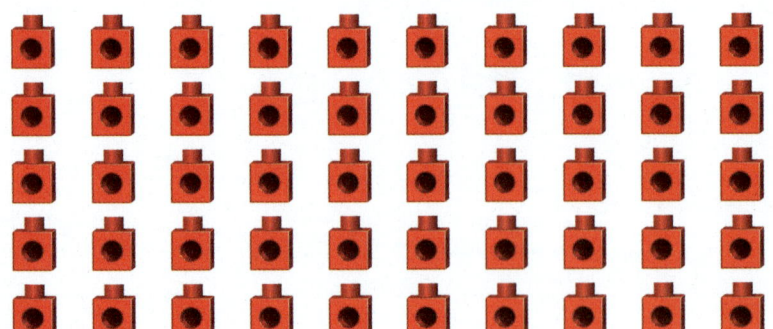

_____ tens and _____ ones is _____.

350   three hundred fifty

**Tens and Ones**

9.

Tens	Ones
(9 tens shown)	(9 ones shown)

→

Tens	Ones

_____ tens and _____ ones is _____.

10. **Modeling Real Life** You have 6 boxes of plastic cups and 3 extra cups. Each box has 10 cups. How many cups are there in all?

_____ plastic cups

**Make Quick Sketches**

Make a quick sketch. Complete the sentence.

11. 17

_____ is _____ ten and _____ ones.

12. 84

_____ is _____ tens and _____ ones.

Chapter 6 three hundred fifty-one 351

 **Understand Place Value**

13. 39

_____ tens is _____.

_____ ones is _____.

_____ tens and _____ ones is _____.

Write Numbers in Different Ways

14. Model 59 two ways.

Tens	Ones

_____ tens and _____ ones is 59.

Tens	Ones

_____ tens and _____ ones is 59.

 **Count and Write Numbers to 120**

15.

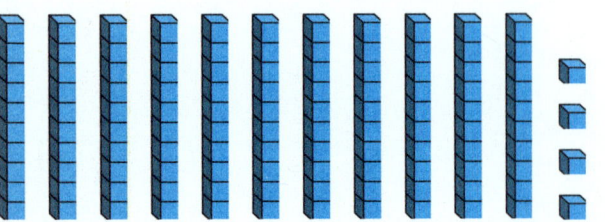

_____

16.

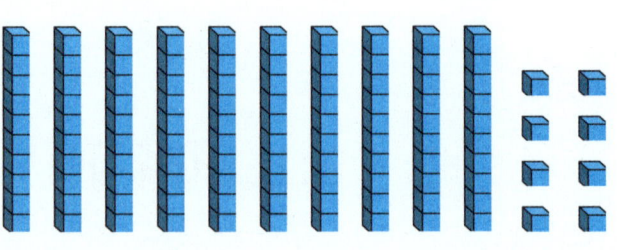

_____

352   three hundred fifty-two

7 Compare Two-Digit Numbers

- What are your favorite toys?
- How many red blocks are there? How many blue blocks are there? Are there more red blocks or blue blocks?

Chapter Learning Target:
Understand two-digit numbers.

Chapter Success Criteria:
- I can identify two-digit numbers.
- I can describe two-digit numbers.
- I can locate two-digit numbers on a number line.
- I can compare two-digit numbers.

Name _____

7 Vocabulary

Review Words
fewer
more

Organize It

Use the review words to complete the graphic organizer.

Define It

Use your vocabulary cards to complete the puzzle.

Across

1. 26 > 23

Down

2.

3. 22 < 38

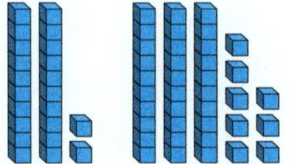

354 three hundred fifty-four

Chapter 7 Vocabulary Cards

compare

greater than

less than

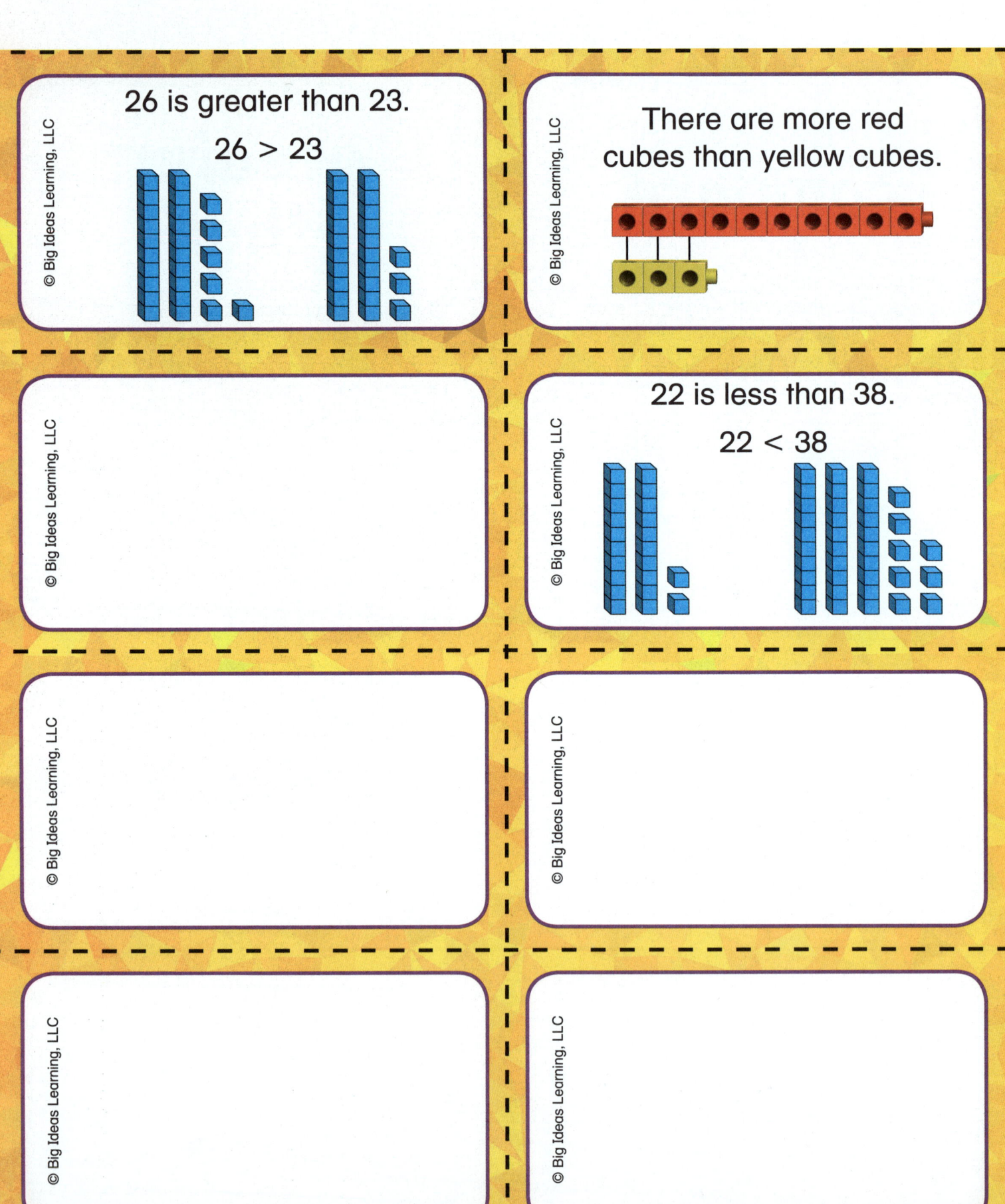

Name _____

Learning Target: Compare two numbers between 11 and 19.

Compare Numbers 11 to 19 — 7.1

Explore and Grow

Model each number. Circle the greater number.

15

17

Chapter 7 | Lesson 1   three hundred fifty-five   355

Think and Grow

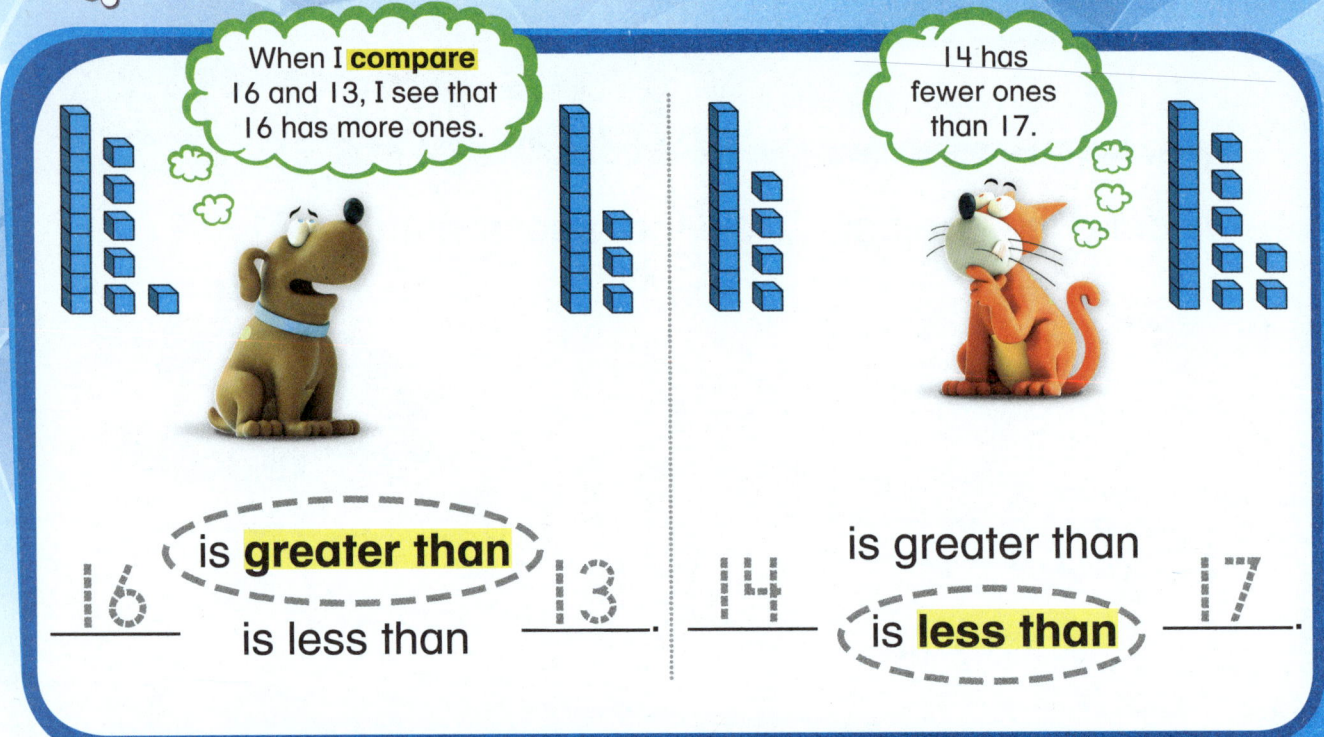

Show and Grow I can do it!

1.

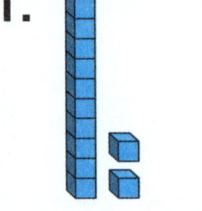

_____ is greater than _____.

_____ is less than _____.

2.

_____ is greater than _____.

_____ is less than _____.

3.

_____ is greater than _____.

_____ is less than _____.

4.

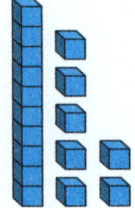

_____ is greater than _____.

_____ is less than _____.

Name _____

 **Apply and Grow: Practice**

5.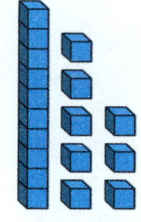

_____ is greater than _____.

_____ is less than _____.

6.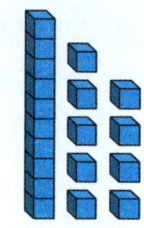

_____ is greater than _____.

_____ is less than _____.

Make quick sketches to compare the numbers.

7.

11 is greater than 12.

11 is less than 12.

8.

15 is greater than 13.

15 is less than 13.

9. **DIG DEEPER!** Choose two numbers to complete the sentences.

12   18
  15
13   19
  16

_____ is greater than _____.

_____ is less than _____.

Chapter 7 | Lesson 1   three hundred fifty-seven   357

Think and Grow: Modeling Real Life

You have 16 tickets. Your friend has 11 tickets and wins 8 more. Who has more tickets?

                    <u>You</u>   <u>Friend</u>

Number of tickets:

Models:

Compare: _____ is greater than _____.

Who has more tickets?   You   Friend

Show and Grow   I can think deeper!

10. You have 7 feathers and find 6 more. Your friend has 12 feathers. Who has more feathers?

                    <u>You</u>   <u>Friend</u>

Number of feathers:

Models:

Compare: _____ is greater than _____.

Who has more feathers?   You   Friend

Name _____

Practice 7.1

Learning Target: Compare two numbers between 11 and 19.

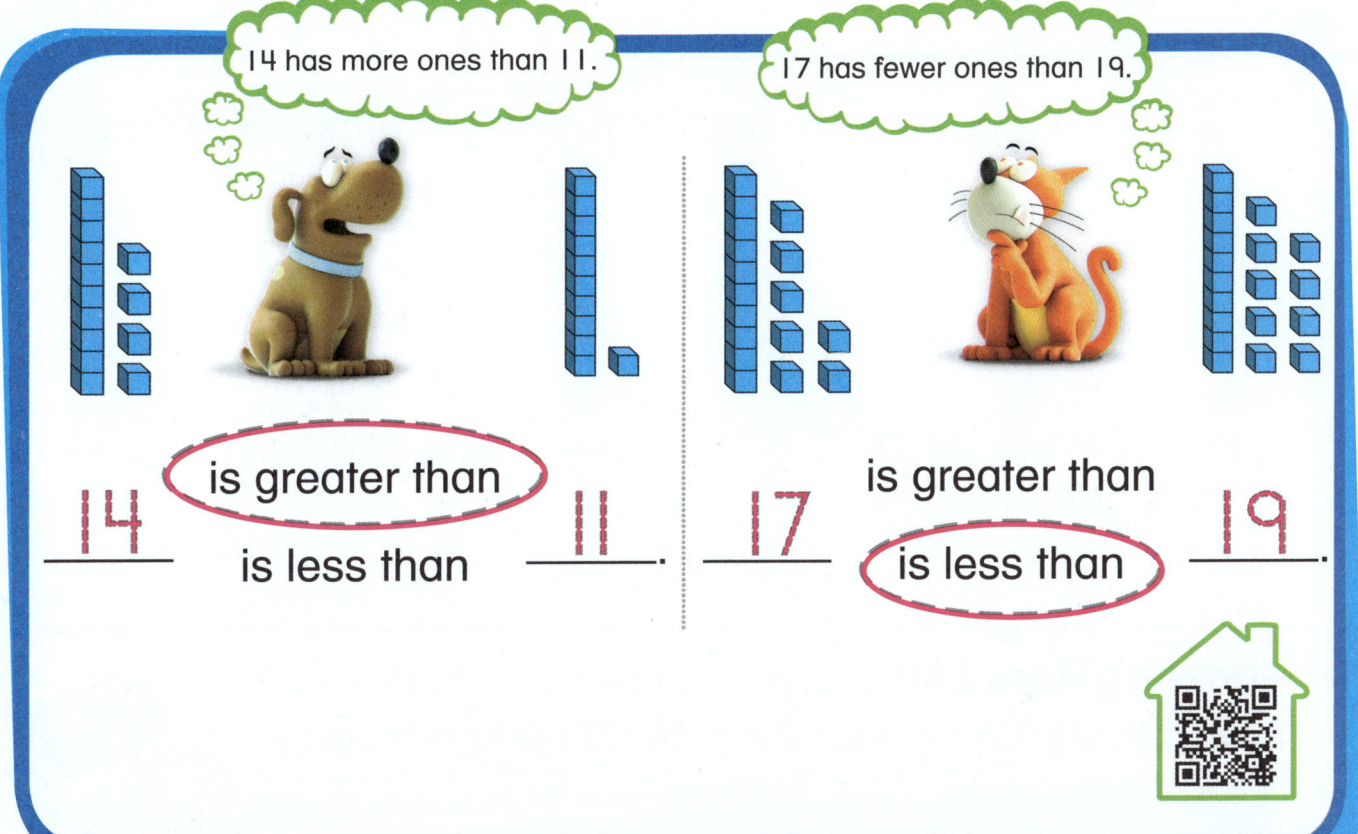

14 has more ones than 11.

17 has fewer ones than 19.

14 (is greater than) / is less than 11.

17 is greater than / (is less than) 19.

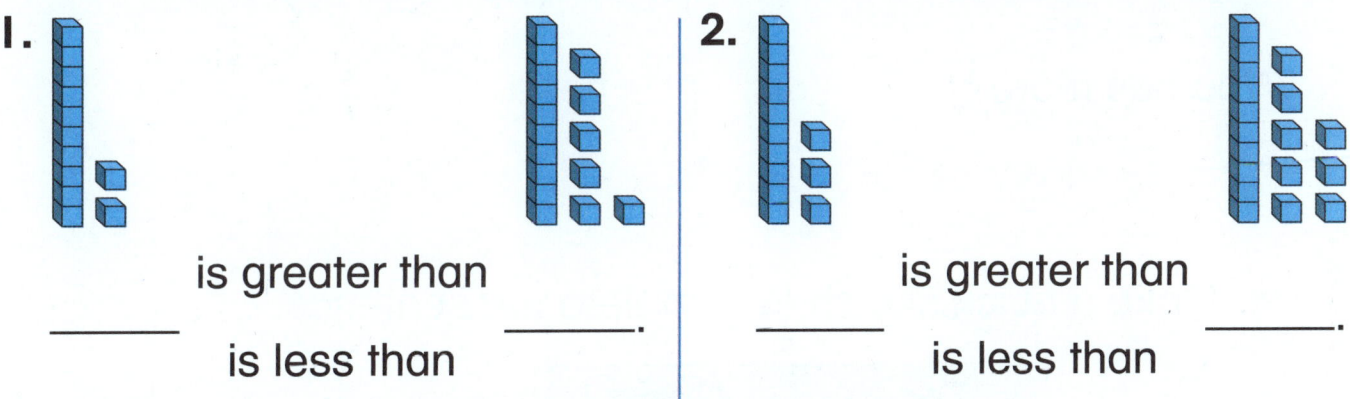

1. _____ is greater than / is less than _____.

2. _____ is greater than / is less than _____.

Make quick sketches to compare the numbers.

3. 19 is greater than / is less than 12.

4. 11 is greater than / is less than 17.

Chapter 7 | Lesson 1   three hundred fifty-nine   359

5. ___ is greater than 13.
17
___ is less than

6. ___ is greater than 14.
12
___ is less than

7. DIG DEEPER! Choose two numbers to complete the sentences.

11   14   17
12   15   18

___ is greater than ___.

___ is less than ___.

8. Modeling Real Life Your tower has 16 red blocks. Your friend's tower has 10 red blocks and 4 blue blocks. Who uses more blocks?

___ is greater than ___.

Who has more?   You   Friend

Review & Refresh

9. Make a quick sketch to complete the sentence.

59

Tens	Ones

___ is ___ tens and ___ ones.

360   three hundred sixty

Name _____

Learning Target: Compare two numbers within 100.

Compare Numbers 7.2

Explore and Grow

Model each number. Circle the greater number.

34

26

Chapter 7 | Lesson 2

three hundred sixty-one 361

Think and Grow

First, compare the tens. 3 tens are fewer than 4 tens.

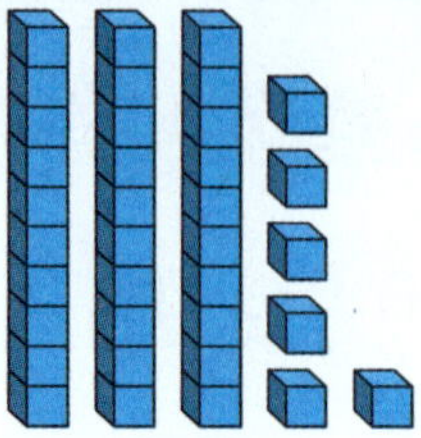

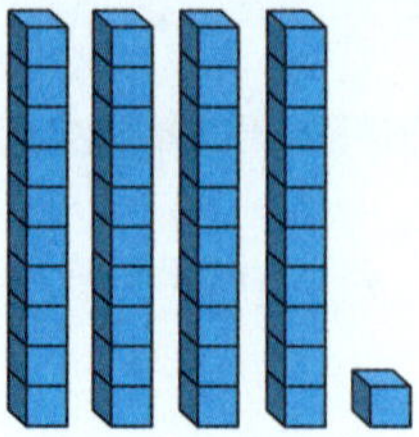

36 is greater than / is less than 41.

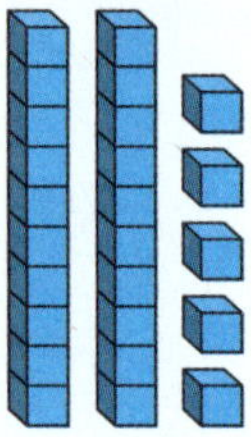

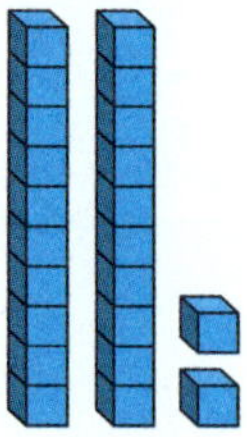

The tens are the same. Compare the ones. 5 ones are more than 2 ones.

25 is greater than / is less than 22.

Show and Grow — I can do it!

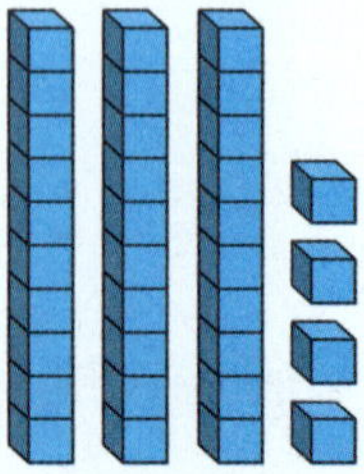

1. 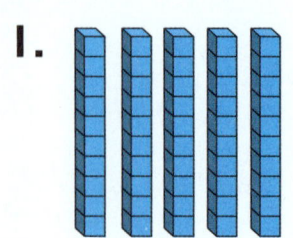 ___ is greater than / is less than ___.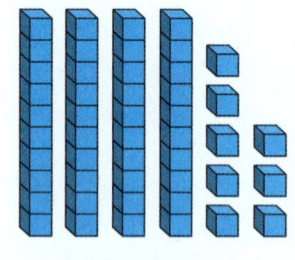

2. 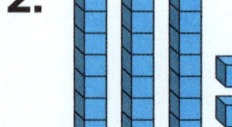 ___ is greater than / is less than ___.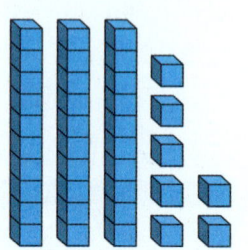

Apply and Grow: Practice

3. 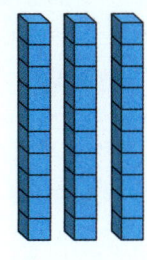 ___ is greater than ___.
___ is less than ___.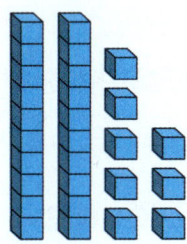

4. 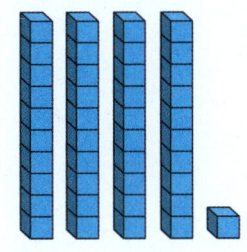 ___ is greater than ___.
___ is less than ___.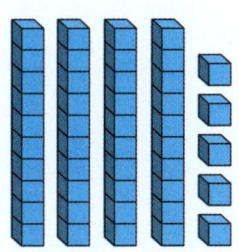

5. Make quick sketches to compare the numbers.

63 is greater than 80.
___ is less than ___.

6. **DIG DEEPER!** Write a number that is greater than 90 but less than 94. Show how you know.

___

Chapter 7 | Lesson 2

Think and Grow: Modeling Real Life

Newton collects 61 acorns. Descartes collects 75 acorns. Who collects more acorns?

Models: Newton   Descartes

Compare: _____ is greater than _____.

Who collects more acorns?   Newton   Descartes

Show and Grow   I can think deeper!

7. You pick 57 blueberries. Your friend picks 53 blueberries. Who picks more blueberries?

    Models: You   Friend

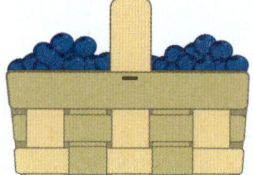

Compare: _____ is greater than _____.

Who picks more blueberries?   You   Friend

Name _____

Practice 7.2

Learning Target: Compare two numbers within 100.

First, compare the tens. 4 tens are more than 2 tens.

<u>45</u> (is greater than) / is less than <u>22</u>.

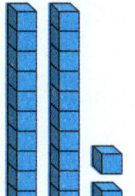

1. 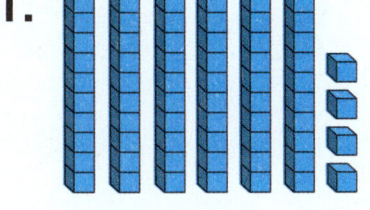 ____ is greater than / is less than ____ .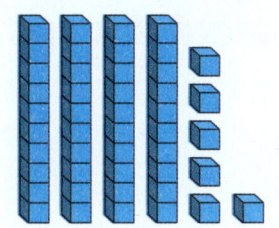

2. 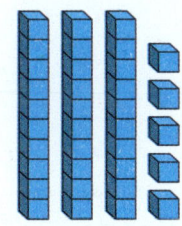 ____ is greater than / is less than ____ .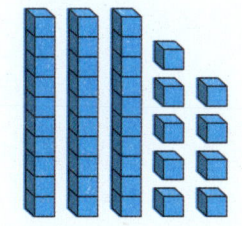

3. Make quick sketches to compare the numbers.

   91   is greater than / is less than   70.

Chapter 7 | Lesson 2   three hundred sixty-five   365

4. 68 _____ is greater than 86.
   _____ is less than

5. **Number Sense** Choose 2 numbers between 50 and 99. Write a sentence to compare the numbers.

   _____   _____

6. **Modeling Real Life** You collect 37 stamps. Your friend collects 27 stamps. Who collects more stamps?

   Who collects more stamps?   You   Friend

Review & Refresh

Circle the taller object.

7.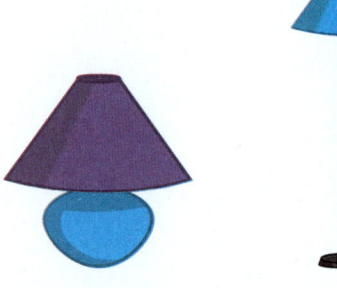

8.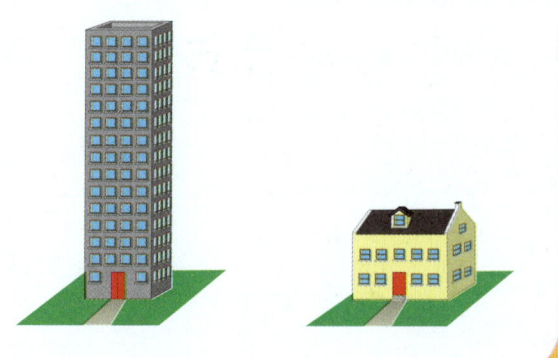

Name _____

Learning Target: Use place value to compare two numbers within 100.

Compare Numbers Using Place Value 7.3

Explore and Grow

Model each number. What is the same about the models? What is different? Circle the greater number.

32

Tens	Ones

23

Tens	Ones

Chapter 7 | Lesson 3

three hundred sixty-seven 367

Think and Grow

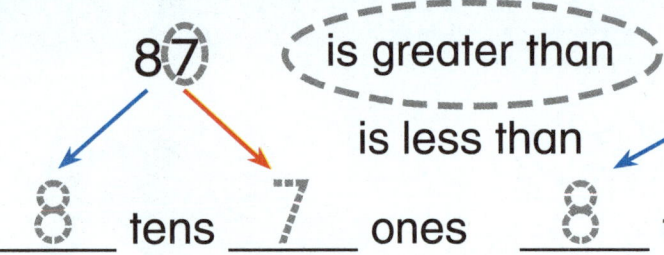

87 **is greater than** 83.
is less than

__8__ tens __7__ ones   __8__ tens __3__ ones

> The tens digits are the same. The ones digits help me decide.

> 2 tens are less than 4 tens. The tens digits help me decide.

29 is greater than 42.
is less than

__2__ tens __9__ ones   __4__ tens __2__ ones

Show and Grow I can do it!

Compare. Which digits help you decide?

1. 61 is greater than 53.
   is less than

   _____ tens _____ one   _____ tens _____ ones

2. 70 is greater than 74.
   is less than

   _____ tens _____ ones   _____ tens _____ ones

Name _____

 **Apply and Grow: Practice**

Compare. Which digits help you decide?

3. _____ 39 is greater than _____ 48.
    is less than

    ____ tens ____ ones   ____ tens ____ ones

4. _____ 80 is greater than _____ 62.
    is less than

    ____ tens ____ ones   ____ tens ____ ones

5. _____ 26 is greater than _____ 23.
    is less than

6. _____ 51 is greater than _____ 86.
    is less than

7. _____ 17 is greater than _____ 71.
    is less than

8. _____ 97 is greater than _____ 92.
    is less than

9. **Precision** Match each ball with its bucket.

Less than 65

58 67 62
64 73 68

Greater than 65

Chapter 7 | Lesson 3   three hundred sixty-nine 369

Think and Grow: Modeling Real Life

Who has more points?

Newton's points: _____   Descartes's points: _____

Compare: _____ is greater than _____.

_____ has more points.

Show and Grow — I can think deeper!

10. Who has more points?

Newton's points: _____   Descartes's points: _____

Compare: _____ is greater than _____.

_____ has more points.

Name _____

Practice 7.3

Learning Target: Use place value to compare two numbers within 100.

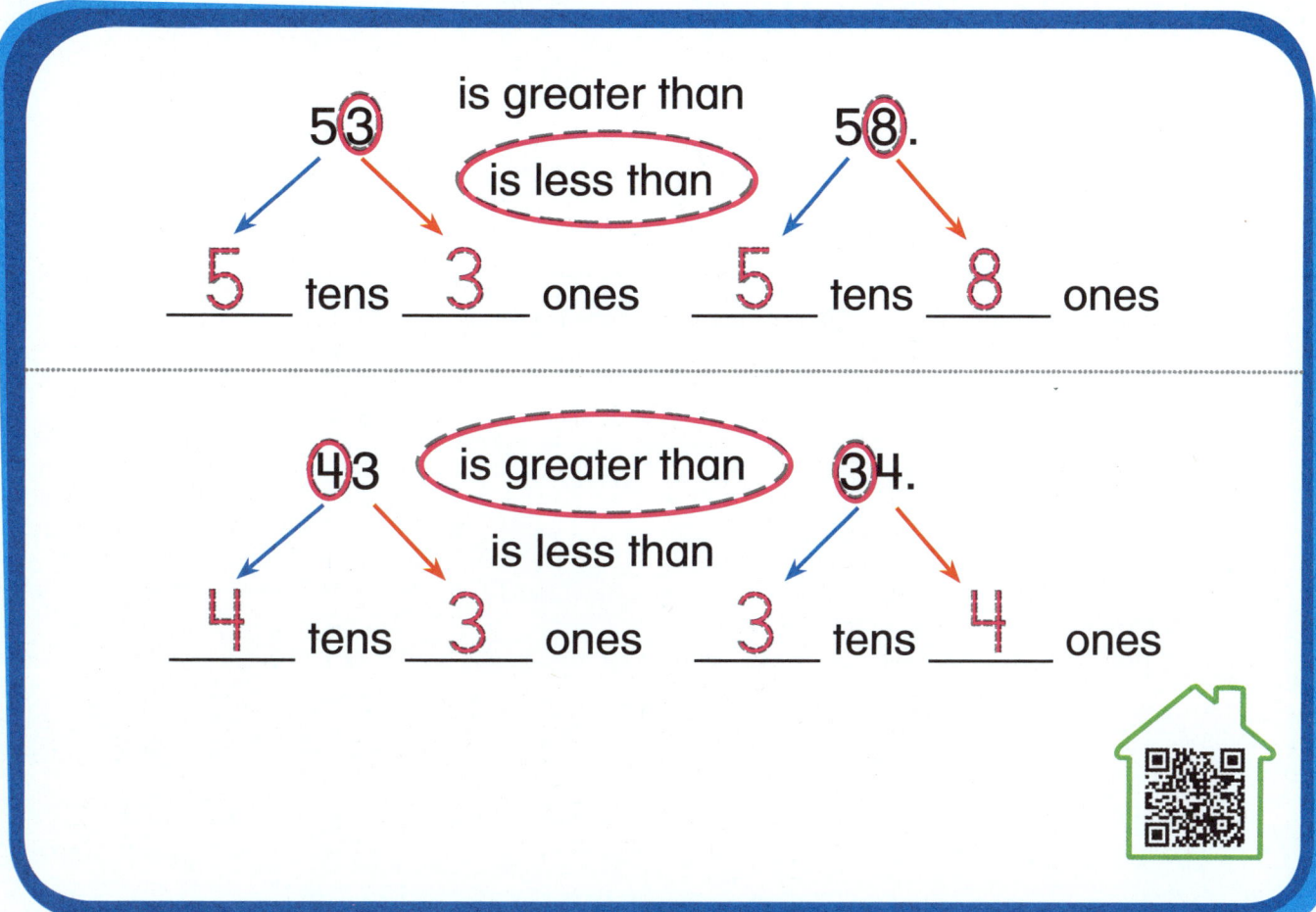

Compare. Which digits help you decide?

1. 16 is greater than / is less than 12.

    _____ ten _____ ones   _____ ten _____ ones

2. 32 is greater than / is less than 38.

3. 7 is greater than / is less than 25.

Chapter 7 | Lesson 3   three hundred seventy-one 371

4.  **Precision** Match each card with its pile.

| Less than 85 | 93 79 84 | Greater than 85 |
| | 81 87 89 | |

5. **Modeling Real Life** Who earns more points?

You

10 Points   1 Point

Friend
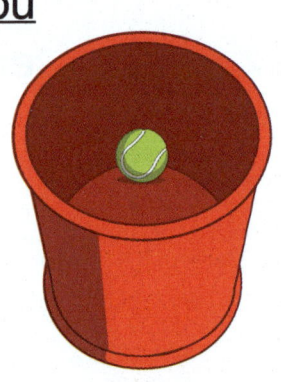
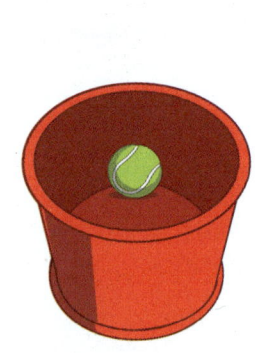
10 Points   1 Point

_____ is greater than _____.

Who earns more points?   You   Friend

Review & Refresh

Is the equation true or false?

6. $15 - 9 \stackrel{?}{=} 11 - 5$

   _____ $\stackrel{?}{=}$ _____

   True   False

7. $4 + 7 \stackrel{?}{=} 16 - 4$

   _____ $\stackrel{?}{=}$ _____

   True   False

372   three hundred seventy-two

Name _____

Learning Target: Use symbols to compare two numbers within 100.

Compare Numbers Using Symbols 7.4

Explore and Grow

Make quick sketches. Complete the sentences.

27

Less than 27	Equal to 27	Greater than 27
____ < 27	____ = 27	____ > 27

Chapter 7 | Lesson 4   three hundred seventy-three   373

Think and Grow

< means *is less than.*
> means *is greater than.*
= means *is equal to.*

Remember: Compare tens first. Then compare ones.

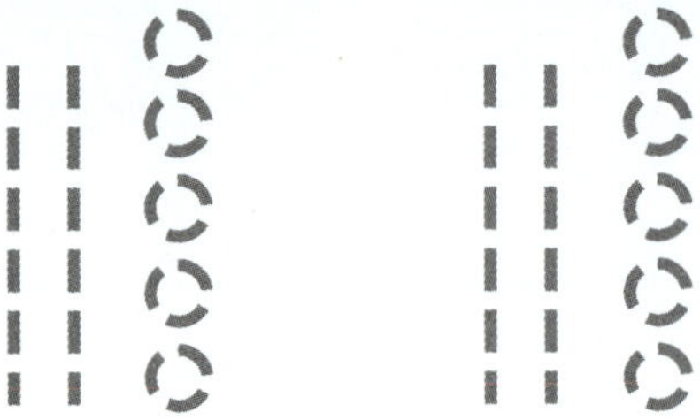

25 __is equal to__ 25.

25 ⬌ 25

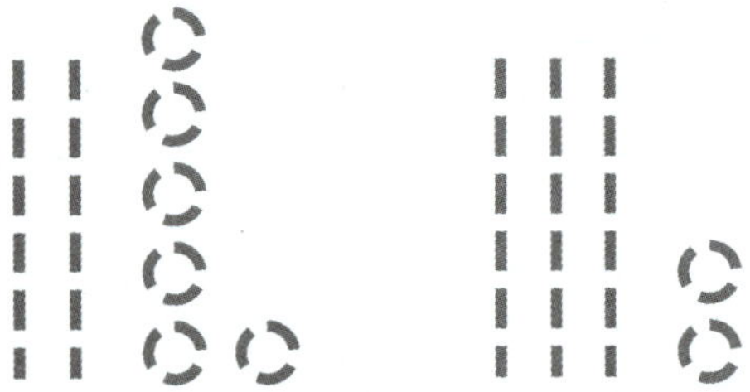

26 __is less than__ 32.

26 < 32

41 __is greater than__ 14.

41 > 14

Show and Grow — I can do it!

Make quick sketches to model each number. Compare.

1. 53 _____ 39.

   53 ◯ 39

2. 28 _____ 34.

   28 ◯ 34

374   three hundred seventy-four

Name _____

 **Apply and Grow: Practice**

Make quick sketches to model each number. Compare.

3.

61 _____ 72.

61 ◯ 72

4.

47 _____ 47.

47 ◯ 47

Compare.

5.

83 _____ 49.

83 ◯ 49

6.

75 _____ 99.

75 ◯ 99

7. 54 ◯ 70

8. 17 ◯ 9

9. 86 ◯ 86

10. **DIG DEEPER!** Choose 2 cards for each problem. Compare the numbers. Use each card once.

31   24   89   13   24   63

____ < ____   ____ > ____   ____ = ____

Chapter 7 | Lesson 4   three hundred seventy-five   **375**

 **Think and Grow: Modeling Real Life**

You have 90 beads. Your friend has 75 beads. Who has more beads?

Models:

Compare: _____ ◯ _____

Who has more beads?   You   Friend

Show and Grow — I can think deeper!

11. You have 48 toy figures. Your friend has 54 toy figures. Who has fewer toy figures?

Models:

Compare: _____ ◯ _____

Who has fewer toy figures?   You   Friend

Name _____

Practice 7.4

Learning Target: Use symbols to compare two numbers within 100.

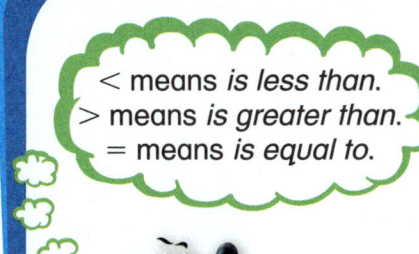

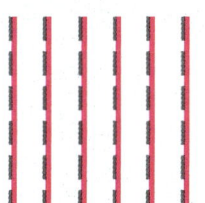

44 __is less than__ 60.

44 < 60

Make quick sketches to model each number. Compare.

1.

37 _____ 15.

37 ◯ 15

2.

22 _____ 22.

22 ◯ 22

Compare.

3. 97 ◯ 79

4. 51 ◯ 83

5. 39 ◯ 62

Chapter 7 | Lesson 4

Compare.

6. 19 ◯ 91

7. 73 ◯ 68

8. 32 ◯ 32

9. **DIG DEEPER!** Use each of the numbers once to complete the puzzle.

74   45   21

81 > _____   56 < _____   21 = _____

10. **Modeling Real Life** Who has more points?

_____ ◯ _____

_____ has more points.

Review & Refresh

11. 6 + 8 = _____

12. 12 + 5 = _____

13. _____ = 0 + 11

14. _____ = 4 + 9

Name _____

Learning Target: Use a number line to compare two numbers within 100.

Compare Numbers Using a Number Line 7.5

Explore and Grow

Circle a number that is less than 45. Underline a number that is greater than 45. How do you know you are correct?

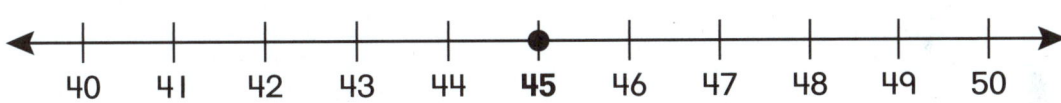

_____ < 45

_____ > 45

Chapter 7 | Lesson 5

three hundred seventy-nine 379

Think and Grow

Numbers to the left of 65 on a number line are less than 65.

Numbers to the right of 65 on a number line are greater than 65.

60 ◯< 65   65 ◯= 65   70 ◯> 65

Show and Grow — I can do it!

Compare.

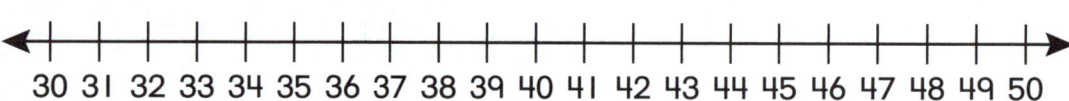

1. 43 ◯ 48

2. 44 ◯ 36

3. 39 ◯ 39

4. 31 ◯ 50

5. 37 ◯ 33

6. 38 ◯ 42

Name _____

✓ Apply and Grow: Practice

Compare.

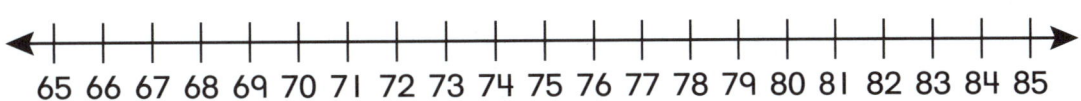

7. 65 ◯ 66

8. 71 ◯ 81

9. 83 ◯ 85

10. 74 ◯ 69

11. 78 ◯ 77

12. 72 ◯ 72

Write a number that makes the statement true.

13. _____ > 47

14. _____ < 76

15. 81 = _____

16.  Newton is thinking of a number that is less than 83 and greater than 74. His number has 6 ones. What is Newton's number?

_____

Think and Grow: Modeling Real Life

The number on your bus is less than 91. Which buses can be yours?

88 90 92

Show how you know:

Show and Grow   I can think deeper!

17. The number on your plane is greater than 58. Which planes can be yours?

63 49 60

Show how you know:

Name _____

Practice  7.5

Learning Target: Use a number line to compare two numbers within 100.

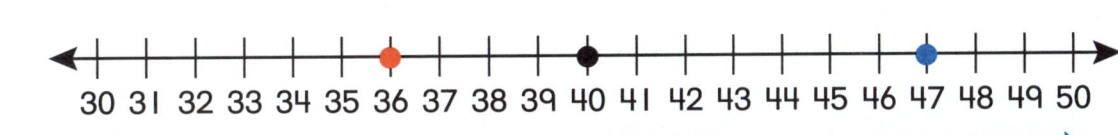

36 < 40   40 = 40   47 > 40

Compare.

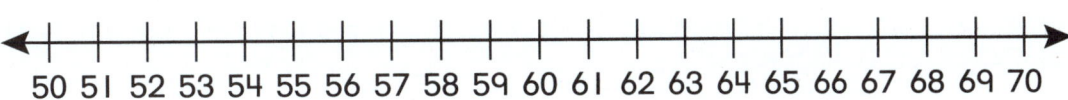

1. 53 ◯ 56

2. 70 ◯ 60

3. 63 ◯ 67

4. 68 ◯ 55

Write a number that makes the statement true.

5. 55 < ____

6. 62 = ____

7. 75 > ____

Chapter 7 | Lesson 5 three hundred eighty-three **383**

8. **Structure** How can you use a number line to tell whether 68 is greater than or less than 42?

9. **Modeling Real Life** The number on your taxi is greater than 40. Which taxis can be yours?

39

46

51

Show how you know:

Review & Refresh

10. You have 9 pencils. 7 are blue. The rest are orange. How many orange pencils do you have?

_____ orange pencils

Name _____

I More, I Less; 10 More, 10 Less 7.6

Learning Target: Identify numbers that are 1 more, 1 less, 10 more, and 10 less than a number.

Explore and Grow

Model 43. Use your model to complete the sentences.

1 more than 43 is _____.

1 less than 43 is _____.

10 more than 43 is _____.

10 less than 43 is _____.

Chapter 7 | Lesson 6

 **Think and Grow**

24

1 more than 24 is __25__.

1 less than 24 is __23__.

10 more than 24 is __34__.

10 less than 24 is __14__.

Show and Grow  I can do it!

1.

1 more than 47 is ____.

1 less than 47 is ____.

10 more than 47 is ____.

10 less than 47 is ____.

2.

1 more than 61 is ____.

1 less than 61 is ____.

10 more than 61 is ____.

10 less than 61 is ____.

Name _____

 **Apply and Grow: Practice**

3.

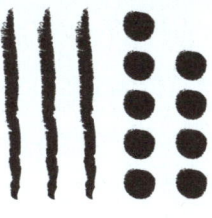

1 more than 39 is _____.   10 more than 39 is _____.

1 less than 39 is _____.   10 less than 39 is _____.

	1 more	1 less	10 more	10 less
4. 56				
5. 75				
6. 33				
7. 80				
8. 12				

9. **Number Sense** Make a quick sketch for the number that is 10 less than the model. What is the new number?

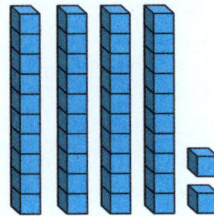

_____

Chapter 7 | Lesson 6   three hundred eighty-seven   387

Think and Grow: Modeling Real Life

You have 25 markers. Newton has 10 more than you. Descartes has 1 fewer than Newton. How many markers does Descartes have?

Models:

_____ markers

Show and Grow   I can think deeper!

10. You have 42 party blowers. Descartes has 10 fewer than you. Newton has 1 more than Descartes. How many party blowers does Newton have?

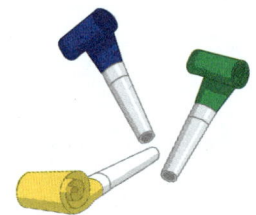

Models:

_____ party blowers

Name _____

Practice 7.6

Learning Target: Identify numbers that are 1 more, 1 less, 10 more, and 10 less than a number.

18

1 more than 18 is __19__.   1 less than 18 is __17__.

10 more than 18 is __28__.   10 less than 18 is __8__.

1.

1 more than 63 is _____.

1 less than 63 is _____.

10 more than 63 is _____.

10 less than 63 is _____.

2.

1 more than 80 is _____.

1 less than 80 is _____.

10 more than 80 is _____.

10 less than 80 is _____.

Chapter 7 | Lesson 6

		1 more	1 less	10 more	10 less
3.	82				
4.	16				
5.	68				

6. **Number Sense** Make a quick sketch for the number that is 1 more than the model. What is the new number?

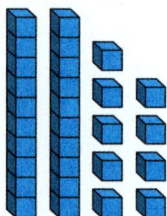

_____

7. **Modeling Real Life** You have 75 party cups. Descartes has 10 more than you. Newton has 1 fewer than Descartes. How many party cups does Newton have?

_____ party cups

Review & Refresh

8. You have 17 erasers. Your friend takes some of them. You have 9 left. How many erasers did your friend take?

_____ erasers

Name _____

Performance Task 7

1. Your school is having a toy drive. Each class wants to collect more than 100 toys.

   a. Your friend's class collects 89 toys. Your cousin's class collects 72 toys. Whose class collects more toys?

   Your _____'s class collects more toys.

   b. Your class collects 10 more toys than your friend's class. Does your class reach the goal?

   Yes   No

2. Use the clues to match each class with the number of toys it collects.

   • Class A collects 10 fewer toys than Class C.
   • Class B collects the fewest number of toys.

   Class A   Class B   Class C

   68   78   88

Number Boss

To Play: Place Number Cards 0–9 in a pile. Each player flips two cards and makes a two-digit number. Compare the numbers. The player with the greater number takes both sets of cards. If the numbers are equal, flip cards again. The player with the greater number takes all of the cards. Repeat until all of the cards have been used.

Name _____

Chapter 7 Practice

7.1 Compare Numbers 11 to 19

1.

   _____ is greater than _____.

   _____ is less than _____.

2.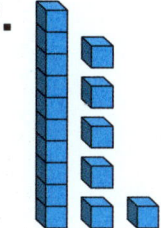

   _____ is greater than _____.

   _____ is less than _____.

7.2 Compare Numbers

3.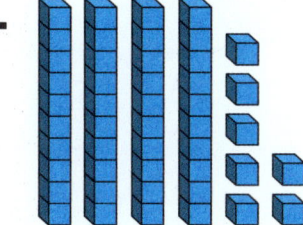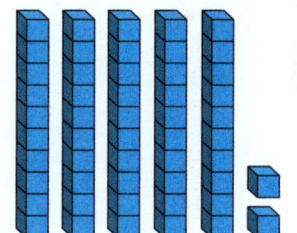

   _____ is greater than _____.

   _____ is less than _____.

4.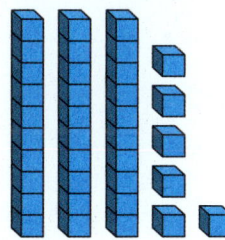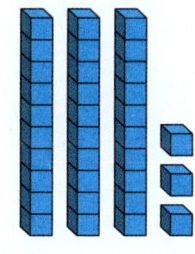

   _____ is greater than _____.

   _____ is less than _____.

5. **MP Structure** Write a number that is greater than 78 but less than 82. Show how you know.

   _____

Chapter 7   three hundred ninety-three   393

7.3 Compare Numbers Using Place Value

Compare. Which digits help you decide?

6.   46   is greater than / is less than   55.

_____ tens _____ ones   _____ tens _____ ones

7.   89   is greater than / is less than   98.

_____ tens _____ ones   _____ tens _____ ones

8. MP Precision Match each chip with its box.

7.4 Compare Numbers Using Symbols

Make quick sketches to model each number. Compare.

9. 84 _____ 68.

84 ◯ 68

10. 29 _____ 42.

29 ◯ 42

7.5 Compare Numbers Using a Number Line

Compare.

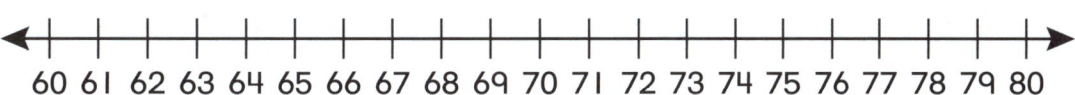

11. 62 ◯ 70

12. 78 ◯ 69

13. 75 ◯ 68

14. 64 ◯ 73

Write a number that makes the statement true.

15. _____ < 36

16. 28 = _____

17. _____ > 9

18. Modeling Real Life The number on your train is less than 34. Which trains can be yours?

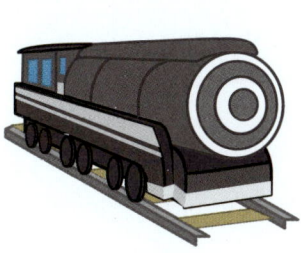
29

42

30

Show how you know:

Chapter 7   three hundred ninety-five

7.6 1 More, 1 Less; 10 More, 10 Less

19.

1 more than 82 is _____.

1 less than 82 is _____.

10 more than 82 is _____.

10 less than 82 is _____.

20.

1 more than 29 is _____.

1 less than 29 is _____.

10 more than 29 is _____.

10 less than 29 is _____.

21. Modeling Real Life You have 34 oranges. Newton has 1 fewer than you. Descartes has 10 fewer than Newton. How many oranges does Descartes have?

_____ oranges

Name _____

Cumulative Practice 1-7

1. 6 people play basketball. Shade the circle next to the picture that shows how many more people need to join the group so there are 10 in all.

○
○
○
○

2. Shade the circles next to the choices that match the model.

○ 6 tens and 3 ones   ○ 63

○ 3 tens and 6 ones   ○ 36

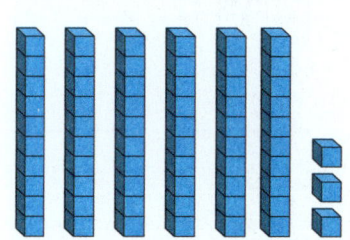

3. Newton has 8 toys. Shade the circle next to the picture that does *not* show a way he could share his toys with Descartes.

○   : 4  : 4   ○  : 7  : 1

○  : 2  : 7   ○  : 3  : 5

Chapter 7   three hundred ninety-seven   397

4. A group of students are at a carnival. 4 of them leave. There are 8 left. How many students were at the carnival to start?

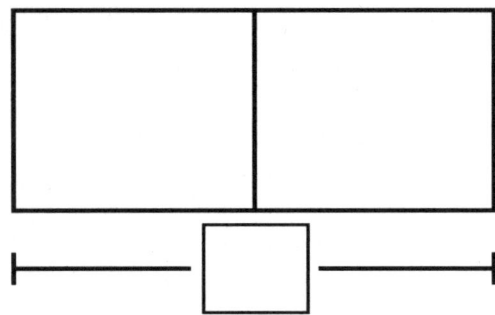

_____ students

---

5. Shade the circle next to the number that completes the puzzle.

 70

 62

 64

○ 73

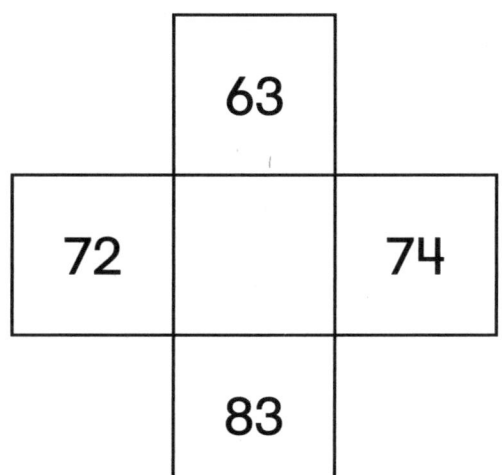

---

6. Circle the two numbers that complete the addition sentence.

2   3   8   9

_____ + _____ + 9 = 19

10

7. Shade the circle next to the number that tells how many ferrets are outside the cage.

○ 5

○ 4

○ 0

○ 1

8. Use the picture to complete the sentence.

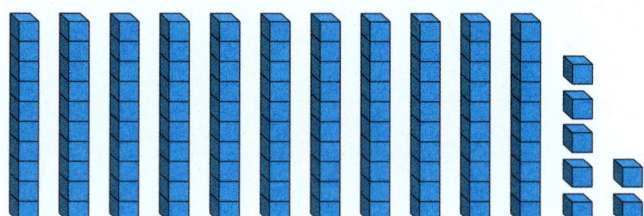

_____ tens and _____ ones is _____.

9. Is each equation true or false?

$0 + 1 \stackrel{?}{=} 0 + 8 + 1$ True   False

$2 + 2 \stackrel{?}{=} 9 - 5$ True   False

$9 - 2 \stackrel{?}{=} 7 - 5$ True   False

$10 - 3 \stackrel{?}{=} 4 + 3$ True   False

10. Write <, >, or = to compare the numbers.

21 ◯ 45   98 ◯ 97

67 ◯ 60   36 ◯ 36

---

11. 4 boats are docked. Some more join them. Now there are 6. Shade the circle next to the models that show how many more boats joined the first group.

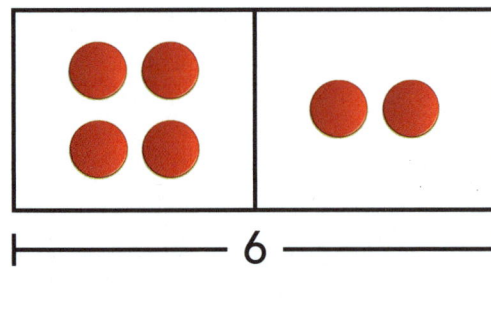

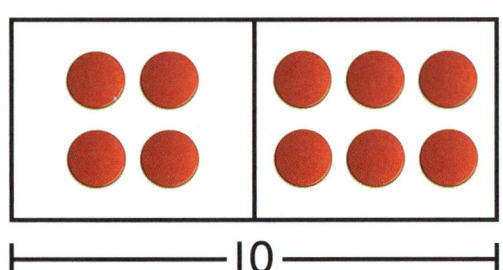

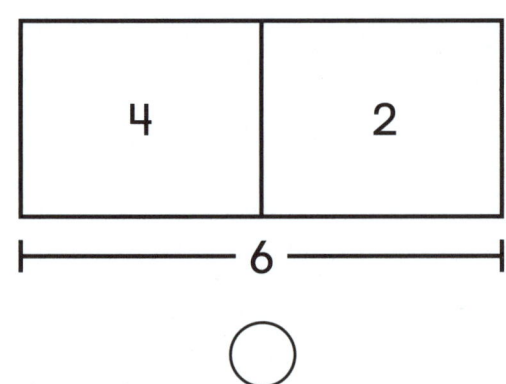

Write two addition equations that can be solved using the double 7 + 7.

____ + ____ = ____   7 + ____ = ____

Glossary

A

add [sumar]

2 + 4 = 6

addend [sumando]

4 + 3 = 7

addition equation
[ecuación de adición]

4 + 5 = 9

analog clock [reloj analogo]

B

bar graph [gráfica de barras]

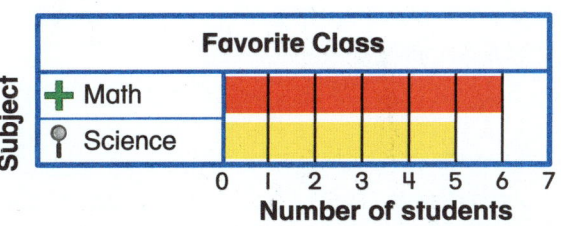

bar model [modelo de barra]

You: | 5 |

Friend: | 2 | 3 |

C

column [columna]

1	2	3	4	5	6	7	8	9	10
11	12	13	14	15	16	17	18	19	20
21	22	23	24	25	26	27	28	29	30
31	32	33	34	35	36	37	38	39	40
41	42	43	44	45	46	47	48	49	50
51	52	53	54	55	56	57	58	59	60
61	62	63	64	65	66	67	68	69	70
71	72	73	74	75	76	77	78	79	80
81	82	83	84	85	86	87	88	89	90
91	92	93	94	95	96	97	98	99	100
101	102	103	104	105	106	107	108	109	110
111	112	113	114	115	116	117	118	119	120

compare   [comparar]

There are more red cubes than yellow cubes.

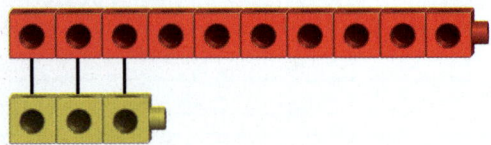

count back   [contar hacia atrás]

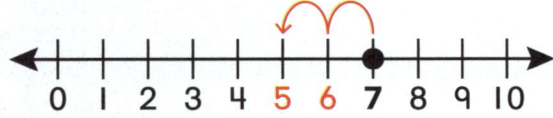

count on   [contar hacia delante]

curved surface   [superficie curva]

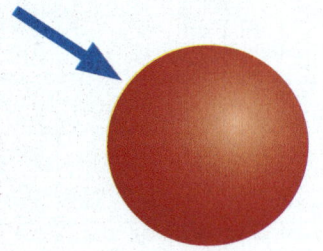

data   [datos]

Favorite Class
math   science
science   math
science   math
math   science
math   science
math

decade numbers
[números de la década]

1	2	3	4	5	6	7	8	9	10
11	12	13	14	15	16	17	18	19	20
21	22	23	24	25	26	27	28	29	30
31	32	33	34	35	36	37	38	39	40
41	42	43	44	45	46	47	48	49	50
51	52	53	54	55	56	57	58	59	60
61	62	63	64	65	66	67	68	69	70
71	72	73	74	75	76	77	78	79	80
81	82	83	84	85	86	87	88	89	90
91	92	93	94	95	96	97	98	99	100
101	102	103	104	105	106	107	108	109	110
111	112	113	114	115	116	117	118	119	120

difference   [diferencia]

$$8 - 3 = 5$$

digit   [dígito]

The digits of 16 are 1 and 6.

16

digital clock [reloj digital]

doubles [dobles]

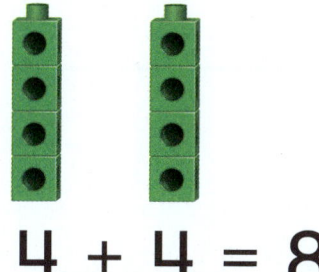

4 + 4 = 8

doubles minus 1
[dobles menos 1]

4 + 4 = 8, so 4 + 3 = 7

doubles plus 1
[dobles más 1]

4 + 4 = 8, so 4 + 5 = 9

edge   [arista]

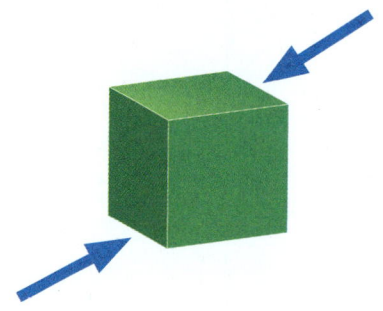

equal shares   [partes iguales]

The squares show **equal shares**.

equals   [igual]

8 + 2 = 10

8 plus 2 equals 10

A3

fact family [hecho de la familia]

2 + 3 = 5
3 + 2 = 5
5 − 2 = 3
5 − 3 = 2

fewer [menos]

flat surface [superficie plana]

fourth of [cuarto de]

A **fourth of** the rectangle is shaded.

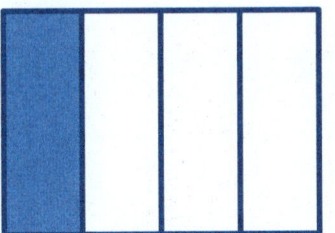

fourths [cuartos]

The rectangle is divided into **fourths**.

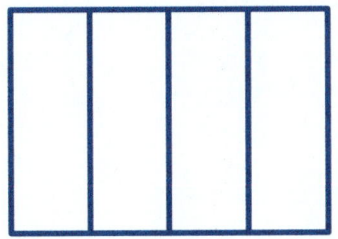

greater than [mayor que]

26 is greater than 23.

26 > 23

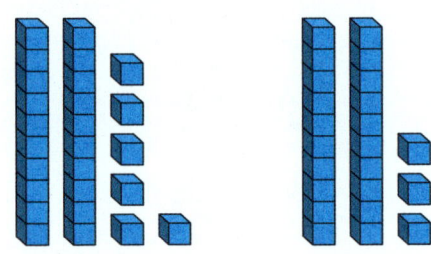

half hour [media hora]

A half hour is 30 minutes.

half of   [mitad de]

Half of the circle is shaded.

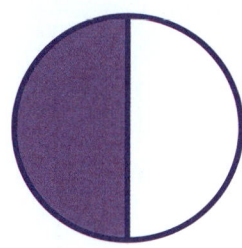

half past   [y media]

half past 3

halves   [mitades]

This circle is divided into **halves**.

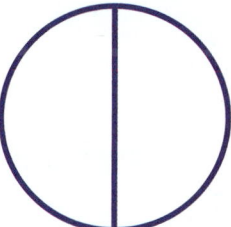

hour   [hora]

An hour is 60 minutes.

hour hand   [horario]

L

length   [longitud]

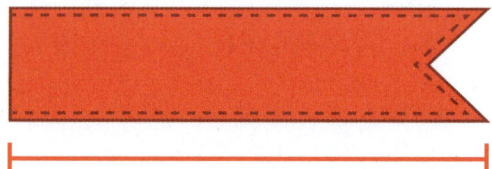

length unit   [unidad de longitud]

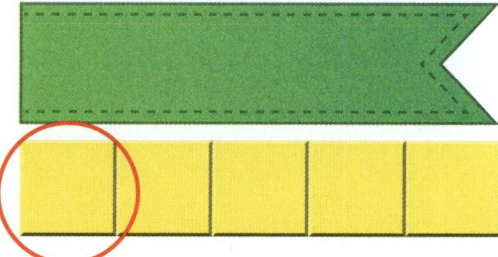

less than   [menor que]

22 is less than 38.
22 < 38

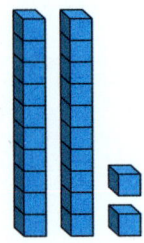 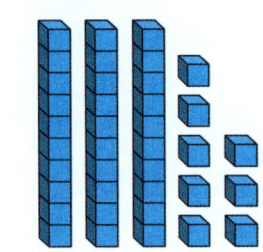

longest   [más largo]

measure   [medida]

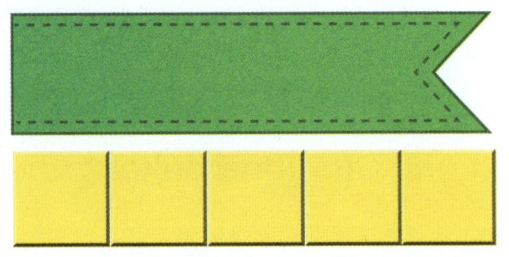

minus   [menos]

3 − 1
3 minus 1

minute   [minuto]

60 minutes is 1 hour.

minute hand   [minutero]

more   [más]

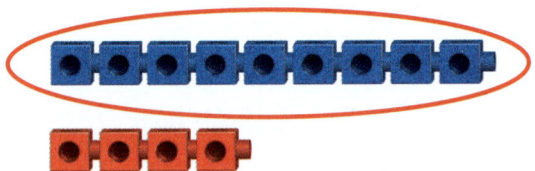

number line   [numero de linea]

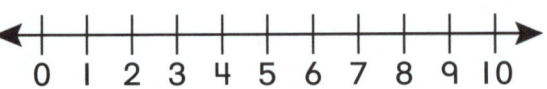

A6

O

o'clock [en punto]

3 o'clock

---

120 chart [120 gráfico]

1	2	3	4	5	6	7	8	9	10
11	12	13	14	15	16	17	18	19	20
21	22	23	24	25	26	27	28	29	30
31	32	33	34	35	36	37	38	39	40
41	42	43	44	45	46	47	48	49	50
51	52	53	54	55	56	57	58	59	60
61	62	63	64	65	66	67	68	69	70
71	72	73	74	75	76	77	78	79	80
81	82	83	84	85	86	87	88	89	90
91	92	93	94	95	96	97	98	99	100
101	102	103	104	105	106	107	108	109	110
111	112	113	114	115	116	117	118	119	120

---

ones [unidades]

23 has 3 ones.

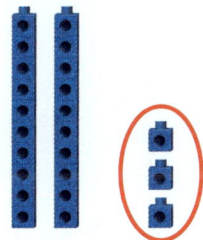

---

ones place [un lugar]

2<u>3</u>

---

open number line [abrir la línea numérica]

P

part [parte]

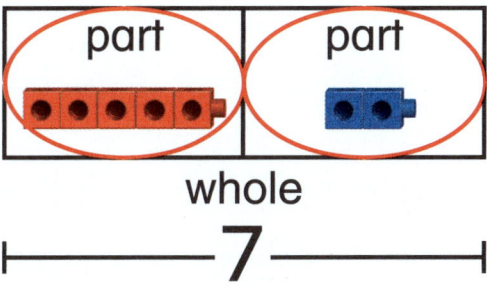

---

part-part-whole model [modelo parte-parte-todo]

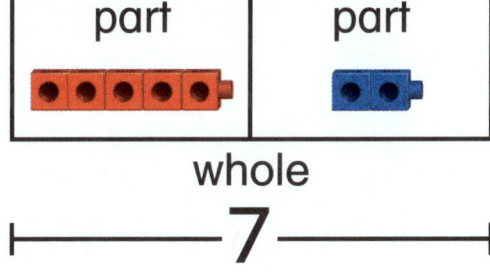

---

picture graph [gráfico de imagen]

Each 🙂 = 1 student.

plus [más]

2 **plus** 1

Q

quarter of [cuarta parte de]

A **quarter of** the rectangle is shaded.

quarters [cuartas partes]

The rectangle is divided into **quarters**.

R

rectangular prism [prisma rectangular]

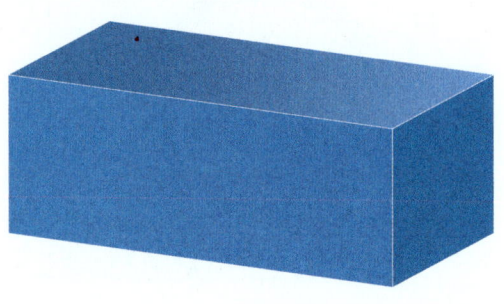

rhombus [rombo]

row [fila]

1	2	3	4	5	6	7	8	9	10
11	12	13	14	15	16	17	18	19	20
21	22	23	24	25	26	27	28	29	30
31	32	33	34	35	36	37	38	39	40
41	42	43	44	45	46	47	48	49	50
51	52	53	54	55	56	57	58	59	60
61	62	63	64	65	66	67	68	69	70
71	72	73	74	75	76	77	78	79	80
81	82	83	84	85	86	87	88	89	90
91	92	93	94	95	96	97	98	99	100
101	102	103	104	105	106	107	108	109	110
111	112	113	114	115	116	117	118	119	120

S

shortest [el más corto]

side [lado]

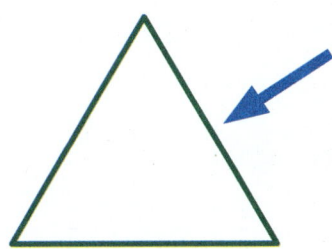

subtract [restar]

$$6 - 4 = 2$$

subtraction equation [ecuación de resta]

$$\underline{9 - 5 = 4}$$

sum [suma]

$$5 + 3 = 8$$

T

tally chart [tabla de conteo]

Favorite Class	
➕ Math	𝍷𝍷𝍷𝍷𝍷 l
💡 Science	𝍷𝍷𝍷𝍷𝍷

tally mark [marca de conteo]

Favorite Class	
➕ Math	𝍷𝍷𝍷𝍷𝍷 ①
💡 Science	𝍷𝍷𝍷𝍷𝍷

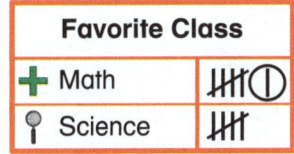

tens [decenas]

23 has 2 tens.

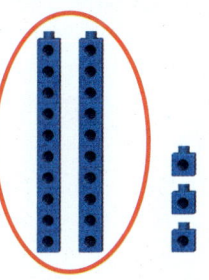

A9

tens place [lugar de decenas]

<u>2</u>3

three-dimensional shape
[forma tridimensional]

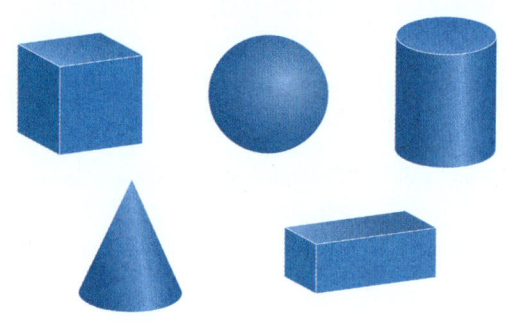

trapezoid [trapecio]

two-dimensional shape
[forma bidimensional]

unequal shares
[partes desiguales]

The shapes show **unequal shares**.

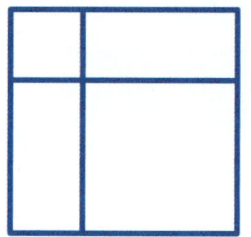 

vertex [vértice]

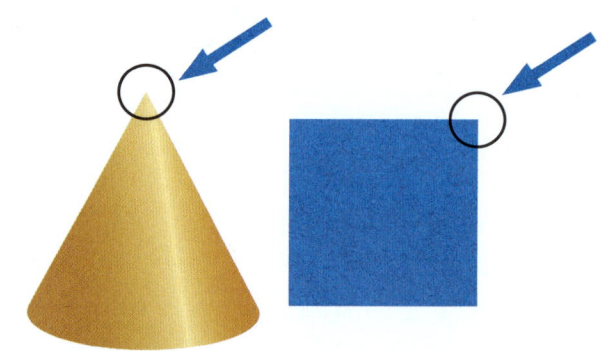

whole [todo]

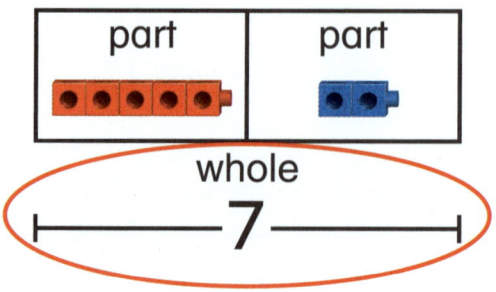

Index

A

Add to
- solving problems with, 3–14
  - with change unknown, 45–50
  - with missing addend, 45–50, 127–132
  - with start unknown, 127–132
- writing equations for, 3–8

Addends, 10, *See also* Addition
- missing or unknown
  - add to problems with, 45–50, 127–132
  - making 10, 163–168
  - part–part–whole model of, 94, 128–132
  - put together problems with, 21–26, 51–56
  - ten frame for finding, 100
- order of, 95–100
- same, in adding doubles, 83–88, 187–192

Addition
- of 0, 65–70
- of 1, 77–82
- in any order, 95–100
- using "count on" strategy, 101–106
  - within 20, 199–204, 230
  - of tens to number, 446, 449
- using doubles, 89–94
  - from 1 to 5, 83–88
  - from 6 to 10, 187–192
  - within 20, 193–198
  - with doubles minus 1 strategy, 89–94, 193–198
  - with doubles plus 1 strategy, 89–94, 193–198
- using "make a 10" strategy
  - adding 9 in, 217–222
  - adding one-digit number to two-digit number in, 471–476
  - adding three numbers in, 211–216
  - adding two numbers in, 223–228
- adding two-digit numbers in, 482
- finding addends in, 163–168
- using part–part–whole model, 16–20
- using place value, 477–482
- practicing strategies of, 483–488
- solving problems using, 489–494
  - within 20, 229–234
  - add to, 3–14
  - add to, with change unknown, 45–50
  - add to, with missing addend, 45–50, 127–132
  - add to, with start unknown, 127–132
  - bigger unknown, 145–150
  - put together, 15–20
  - put together, connected with take apart, 51–56
  - put together, missing both addends, 21–26
- of tens, 415–420
  - using mental math, 403–408
  - to number, 445–450
  - on number line, 421–426, 446, 449
- of tens and ones, 459–464
  - on number line, 465–470, 484, 490, 493
- of three numbers, 205–216

Addition equations, *See also* Addition
- add to, 3–14, 45–50
- completing fact families with, 169–174, 520
- definition of, 4
- put together, 15–26
- true or false, 157–162, 267–272, 372, 482
- writing, 3–8

Addition sentence, 4 (*See also* Addition equations)

Addition to subtract strategy, 113–118
- within 20, 249–254
- of tens, 439–444

All, subtracting, 71–76

Analog clock
    definition of, 584
    using hour and minute hand of, 595–600
    using hour hand of
        to tell time to half hour, 589–594
        to tell time to hour, 583–588
    using to tell time, 601–606
Another Way, 200, 446, 449, 466, 484, 487
Apply and Grow: Practice, *In every lesson. For example, see:* 5, 67, 129, 189, 245, 295, 357, 405, 461, 505

B

Bar graphs
    definition of, 554
    making, 559–564
    picture graphs compared to, 553
    reading and interpreting, 553–558
    solving problems with data from, 565–570
Bar model
    adding within 20 using, 231, 232, 234
    compare problems using
        with bigger unknown, 146–150
        with length measurement, 528–532
        with smaller unknown, 152–156
    definition of, 146
    subtraction using, within 20, 281, 282
Bigger unknown, compare problems with, 145–150

C

Challenge, *See* Dig Deeper
Change, unknown
    add to problems with, 45–50
    take from problems with, 133–138
Chapter Practice, *In every chapter. For example, see:* 59–62, 121–124, 177–180, 237–240, 287–290, 349–352, 393–396, 453–456, 497–500, 535–538

Charts
    120
        for counting by ones, 293–298
        for counting by tens, 299–304
        definition of, 694
    hundred
        for adding 10, 404, 406
        for adding tens and ones, 465–470
        for subtracting 10, 410, 413
    tally
        completing picture and bar graphs with, 559–564
        definition of, 542
        making, 541–546, 559
        picture graphs compared to, 547
        solving problems with data from, 565–570
Choose Tools, 627, 630
Circles
    combining shapes to make, 626, 632
    describing, 619, 621
    as flat surfaces, of shapes, 644–648
    identifying equal shares in, 676, 677, 679
        fourths, 688–692
        halves, 682–686
    taking apart, 639
Clock
    analog and digital, 601–606
    using hour and minute hands of, 595–600
    using hour hand of
        to tell time to half hour, 589–594
        to tell time to hour, 583–588
Closed shapes
    definition of, 614
    sorting, 614–618
Color tiles
    comparing length using, 528–532
    measuring length using, 516–526
    representing data with, 559–564
Columns, in 120 chart, 300
Combining shapes
    three-dimensional, 655–660
    two-dimensional, 625–636

Common Errors, *Throughout. For example, see:* T-22, 199
Common Misconceptions, *Throughout. For example, see:* T-22, T-40, T-102, T-128, T-148
Compare, definition of, 356
Compare problems, solving
    bigger unknown, 145–150
    how many fewer, 39–44, 151–156
    how many more, 33–38, 145–150
    length, 527–532
    smaller unknown, 151–156
Comparing length
    indirectly, using third object, 509–514
    ordering objects by length, 503–508
    solving compare problems, 527–532
Comparing numbers
    1 more, 1 less, 385–390
    10 more, 10 less, 385–390
    11 to 19, 355–360
    within 100, 361–366
        using number line, 379–384
        using place value, 367–372
        using symbols, 373–378
    quick sketches for, 357, 359, 365, 373–375, 444
Composing numbers
    11 to 19, 305–310
    to 120, 341–346
    counting by tens and ones for, 317–322
    decade numbers (tens), 311–316
    in different ways, 335–340
    quick sketches for, 323–328, 360
    understanding place value in, 329–334
Cones
    combining with other shapes, 655–657, 659
    definition of, 650
    describing, 649–654
    taking apart shapes containing, 661–666
"Count back" strategy
    definition of, 108
    subtraction using, 107–112, 434, 437
    within 20, 243–248, 280
"Count on" strategy
    addition using, 101–106
    within 20, 199–204, 230
    of tens to number, 446, 449
    definition of, 102
    subtraction using, 250–254, 440, 443
Counting
    to 120 by ones, 293–298
    to 120 by tens, 299–304
    to add 1, 78
    tens and ones, to write numbers, 317–322
Cross-Curricular Connections, *In every lesson. For example, see:* T-13, T-161, T-197, T-365, T-419, T-487, T-525, T-563, T-605, T-685
Cubes
    combining, to make new shapes, 655–660
    definition of, 650
    describing, 649–654
    taking apart shapes containing, 661–666
Cumulative Practice, 181–184, 397–400, 577–580, 697–700
Curved surfaces
    definition of, 644
    sorting shapes by, 644–648
Cylinders
    combining with other shapes, 655–657, 659
    definition of, 650
    describing, 649–654
    taking apart shapes containing, 661–666

D

Data
    bar graphs of
        definition of, 554
        making, 559–564
        picture graphs compared to, 553
        reading and interpreting, 553–558

solving problems with data from, 565–570
definition of, 542
picture graphs of
bar graphs compared to, 553
definition of, 543
making, 559–564
reading and interpreting, 547–552
solving problems with data from, 565–570
tally charts compared to, 547
representing, 559–564
solving problems involving, 565–570
tally charts of, 541–546
completing picture and bar graphs with, 559–564
definition of, 542
making, 541–546, 559
picture graphs compared to, 547
solving problems with data from, 565–570

Decade numbers, 300, 311–316 (*See also* Tens (10))

Decomposing
take apart problems, 52–56
taking apart shapes
three-dimensional, 661–666
two-dimensional, 637–642

Define It, *In every chapter. For example, see:* 2, 64, 126, 186, 242, 292, 354, 402, 458, 502

Differences, 28 (*See also* Subtraction)

Differentiation, *See* Scaffolding Instruction

Dig Deeper, *Throughout. For example, see:* 5, 67, 195, 245, 304, 357, 405, 482, 508, 555

Digit(s)
comparing in two-digit numbers, 367–372
definition of, 330
value in two-digit number, 329–334

Digital clock
definition of, 602
telling time on, 601–606

Doubles
using
within 20, 193–198
to find sum, 89–94, 193–198
adding
from 1 to 5, 83–88
from 6 to 10, 187–192
within 20, 193–198
with doubles minus 1 strategy, 89–94, 193–198
with doubles plus 1 strategy, 89–94, 193–198
definition of, 84

Doubles minus 1, 89–94, 193–198
Doubles plus 1, 89–94, 193–198

E

Edges, of three-dimensional shapes, 650–654

ELL Support, *In every lesson. For example, see:* T-2, T-127, T-235, T-282, T-312, T-385, T-430, T-465, T-586, T-682

Equal shares
definition of, 676
identifying, 675–680
shapes showing fourths, 687–692
shapes showing halves, 681–686

Equals (equal to), 10, 373–378

Equations
addition
add to, 3–14
completing fact families with, 169–174, 520
definition of, 4
put together, 15–26
true or false, 157–162, 267–272, 372, 482
writing, 3–8
subtraction
completing fact families with, 169–174, 520

A14

how many fewer, 40–44
how many more, 34–38
take apart, 52–56
take from, 27–32
true or false, 157–162, 267–272, 372, 482
writing, 28–32
true, finding number making, 273–278
Error Analysis, *See* You Be the Teacher
Explain, *Throughout. For example, see:* 411, 450, 485, 505, 523, 541, 570, 600, 613, 686
Explore and Grow, *In every lesson. For example, see:* 3, 65, 127, 187, 243, 293, 355, 403, 459, 541

F

Fact families
completing, 169–174, 520
definition of, 170
Fewer
1 or 10, identifying numbers with, 385–390
definition of, 40
how many, compare problems solving for, 39–44, 151–156
Flat surfaces
definition of, 644
describing shapes by, 649–654
sorting shapes by, 644–648
Formative Assessment, *Throughout. For example, see:* T-6, T-74, T-202, T-276, T-344, T-406, T-556, T-604, T-646, T-678
Fourth of, 688
Fourths
definition of, 688
identifying shapes showing, 687–692

G

Games, *In every chapter. For example, see:* 58, 120, 176, 236, 286, 348, 392, 452, 496, 534
"Get to 10" strategy, subtraction using, 261–266, 426
subtracting nine in, 255–260
Graphs
bar
definition of, 554
making, 559–564
picture graphs compared to, 553
reading and interpreting, 553–558
solving problems with data from, 565–570
picture
bar graphs compared to, 553
definition of, 543
making, 559–564
reading and interpreting, 547–552
solving problems with data from, 565–570
tally charts compared to, 547
Greater than (>), 356, 373–378 (*See also* Comparing numbers)
Groups of objects, *See also specific operations and problems*
adding to, 3–14
compare problems
how many fewer, 39–44
how many more, 33–38
putting together, 15–26
taking apart, 51–56
taking from, 27–32

H

Half of, 682, 685
Half hour
on analog and digital clocks, 601–606
definition of, 590

telling time to
hour and minute hands for, 595–600
hour hand for, 589–594

Half past
on analog and digital clocks, 601–606
definition of, 590
hour and minute hands showing, 595–600
hour hand showing, 589–594

Halves
definition of, 682
identifying shapes showing, 681–686

Hexagons
combining, to make new shapes, 628, 631
combining shapes to make, 626–629
definition of, 620
describing, 619–624
equal shares in, identifying, 677, 681
halves of, 681
taking apart, 639, 642

Higher Order Thinking, *See* Dig Deeper

Hour
on analog and digital clocks, 601–606
definition of, 584
telling time to
hour and minute hands for, 595–600
hour hand for, 583–588

Hour hand
definition of, 584
using to tell time to half hour, 589–594
using to tell time to hour, 583–588
using to tell time to hour and half hour, 595–600

Hundred chart
for adding 10, 404, 406
for adding tens and ones, 465–470
for subtracting 10, 410, 413

Learning Target, *In every lesson. For example, see:* 3, 65, 127, 187, 243, 293, 355, 459, 503, 541

Length
comparing indirectly, 509–514
measuring
using color tiles, 516–526
using like objects, 515–520
using paper clips, 521–526
ordering objects by, 503–508
solving compare problems involving, 527–532

Length unit
color tiles of, 516–520
definition of, 516

Less than (<), 356, 373–378 (*See also* Comparing numbers)

Linking cubes
for add to problems, 3, 9, 10, 13
for adding 0, 65
for adding 1, 77
for adding doubles, 84–88, 90, 188, 189, 191
for adding in any order, 96, 97, 99
for adding three numbers, 205
for adding within 20, 233
for completing fact families, 169
for composing numbers
11 to 19, 306, 309
decade numbers (tens), 312–316
for doubles minus 1 strategy, 90, 91, 93, 194, 195, 197
for doubles plus 1 strategy, 90, 91, 93, 194, 195, 197
for grouping by 10, 312–316
for put together problems, 21
for take from problems, 27, 28, 31

Logic, 67, 70, 473, 476

Longest
definition of, 504
ordering objects by, 503–508

L-shaped vertices, 615, 616, 619, 622

M

"Make a 10" strategy, addition using
  adding 9 in, 217–222
  adding one-digit number to two-digit number in, 471–476
  adding three numbers in, 211–216
  adding two numbers in, 223–228
  adding two-digit numbers in, 482
  finding addends in, 163–168

Math Musicals, *In every chapter of the Teaching Edition. For example, see:* 4, 84, 108, 218, 300, 516, 596, 682

Mental math
  adding 10 using, 403–408
  subtracting 10 using, 409–414

Minus (−), 28

Minute
  on analog and digital clocks, 601–606
  definition of, 596
  telling time to, 595–600

Minute hand
  definition of, 596
  using to tell time to hour and half hour, 595–600

Missing addends
  add to problems with, 45–50, 127–132
  making 10, 163–168
  part–part–whole model of, 94, 128–132
  put together problems with, 21–26, 51–56
  ten frame for finding, 100

Modeling, of numbers, *See also* Bar model; Part–part–whole model
  in different ways, 335–340
  as tens and ones, 323–328, 360
  two-digit, 329–334

Modeling Real Life, *In every lesson. For example, see:* 8, 70, 132, 192, 248, 304, 360, 408, 464, 508

More
  1 or 10, identifying numbers with, 385–390
  definition of, 34
  how many, compare problems solving for, 33–38, 145–150

Multiple Representations, *Throughout. For example, see:* 4, 103, 170, 218, 260, 306, 351, 446, 547, 692

N

Nine (9)
  adding, using "make a 10" strategy, 217–222
  subtracting, in "get to 10" strategy, 255–260

Number line
  adding on
    within 20, 199–204
    using "count on" strategy, 101–106, 199–204, 230, 446, 449
    solving word problems with, 230, 234
    of tens, 421–426, 446, 449
    of tens and ones, 465–470, 484, 490, 493
  comparing numbers on, 379–384
  definition of, 102
  finding number making true equation on, 273
  open, definition of, 422
  subtracting on
    within 20, 243–254, 280
    using addition to subtract strategy, 249–254
    using "count back" strategy, 107–112, 243–248, 280
    using "count on" strategy, 250–254, 440, 443
    of tens, 433–438

Number Sense, *Throughout. For example, see:* 17, 20, 91, 94, 97, 100, 123, 159, 225, 263, 307, 366, 387, 390

O

120
- counting to, 341–346
  - by ones, 293–298
  - by tens, 299–304
- writing numbers to, 341–346

120 chart
- for counting by ones, 293–298
- for counting by tens, 299–304
- definition of, 294

Objects, groups of, *See also specific operations and problems*
- adding to, 3–14
- compare problems
  - how many fewer, 39–44
  - how many more, 33–38
- putting together, 15–26
- taking apart, 51–56
- taking from, 27–32

O'clock
- on analog and digital clocks, 601–606
- to half hour, 589–594
- to hour, 583–588
- to hour and half hour, 595–600

Ones (1)
- adding, 77–82
  - in adding tens to number, 445–450
  - in adding two numbers, with tens, 459–470
  - in adding two-digit numbers, 477–482
  - on number line, 465–470, 484, 490, 493
  - one-digit numbers to two-digit number, 471–476
- in comparing numbers
  - 1 more, 1 less, 385–390
  - 11 to 19, 355–360
  - within 100, 361–372, 374
- in composing or writing numbers
  - 11 to 19, 305–310
  - to 120, 341–346
  - counting for, 317–322
  - in different ways, 335–340
  - quick sketches for, 323–328, 360
  - in two-digit number, 329–334
- counting to 120 by, 293–298
- definition of, 306
- subtracting, 77–82

Ones place, 306, 329–334

Open number line, *See also* Number line
- definition of, 422

Open shapes
- definition of, 614
- sorting, 614–618

Organize It, *In every chapter. For example, see:* 2, 64, 126, 186, 242, 292, 354, 402, 612, 674

P

Paper clips
- comparing length using, 528–532
- measuring length using, 521–526

Part, definition of, 16

Partitioning shapes
- into equal shares, 675–680
- into fourths, 687–692
- into halves, 681–686

Part–part–whole model
- add to using, 46–50, 128–132
- adding within 20 using, 233
- addition to subtract strategy using, 114–118
- completing fact families using, 170, 171, 173
- definition of, 16
- finding missing addend in, 94, 128–132
- put together using, 16–20
- put together/take apart using, 52–56
- subtracting within 20 using, 281, 283, 284
- take from problems using
  - with change unknown, 134–138
  - with start unknown, 140–144

A18

Performance Task, *In every chapter. For example, see:* 57, 119, 175, 235, 285, 347, 391, 451, 495, 533, 571, 607, 667, 693

Picture graphs
    bar graphs compared to, 553
    definition of, 543
    making, 559–564
    reading and interpreting, 547–552
    solving problems with data from, 565–570
    tally charts compared to, 547

Place value
    in addition of two-digit numbers, 477–482
    in comparing numbers
        11 to 19, 355–360
        within 100, 367–372
    in composing or writing numbers
        11 to 19, 305–310
        to 120, 341–346
        counting tens and ones for, 317–322
        decade numbers (tens), 311–316
        in different ways, 335–340
        quick sketches for, 323–328, 360
        in two-digit numbers, 329–334
    understanding, 329–334

Plus sign (+), 10

Practice, *In every lesson. For example, see:* 7–8, 69–70, 131–132, 191–192, 247–248, 297–298, 359–360, 407–408, 463–464, 507–508

Precision, *Throughout. For example, see:* 29, 147, 369, 372, 517, 585, 588, 621, 624, 680

Prisms, rectangular
    combining, to make new shapes, 655–657, 659
    definition of, 650
    describing, 649–654
    taking apart shapes containing, 661–666

Problem solving, *See* Word problems

Problem Solving Strategy, *Throughout. For example, see:* 230, 233, 282, 490, 492

Problem Types, *Throughout. For example, see:*
    add to
        change unknown, 45, 106, 168, 233, 400, 491
        result unknown, 4, 65, 162, 199, 228, 272, 408, 426, 450, 564
        start unknown, 130, 132, 233
    compare
        bigger unknown, 146, 178, 204, 232, 240, 470, 486, 500
        difference unknown, 34, 119, 180, 246, 285, 451, 528, 576, 658, 697
        smaller unknown, 152, 179, 196, 264, 283, 432, 493, 530, 594
    put together
        addend unknown, 98, 100, 175, 183, 562
        both addends unknown, 22, 57, 60, 86, 192
        total unknown, 16, 52, 119, 187, 205, 235, 418, 462, 480, 579
    take apart
        addend unknown, 116, 175, 183, 562
        both addends unknown, 57, 397
        total unknown, 52, 55, 62, 119, 258
    take from
        change unknown, 136, 138, 252, 279, 290
        result unknown, 28, 71, 112, 243, 412, 436, 451, 526, 567, 578
        start unknown, 142, 281, 290, 398, 700

Put together problems, solving, 15–20
    connected with take apart, 51–56
    missing both addends, 21–26

Quarters
    definition of, 688
    identifying, 687–692

Quick sketches
    for adding tens to number, 446, 449
    for comparing numbers, 357, 359, 365, 373–375, 444
    for modeling numbers as tens and ones, 323–328, 360
    for modeling two-digit numbers, 329–334

R

Reading, *Throughout. For example, see:* T-7, T-87, T-155, T-247, T-297, T-425, T-513, T-551, T-587, T-679
Real World, *See* Modeling Real Life
Reasoning, *Throughout. For example, see:* 122, 153, 189, 461, 520, 543, 624, 639, 642, 692
Rectangles
    combining, to make squares, 632
    combining shapes to make, 626, 627, 633
    definition of, 620
    describing, 619–624
    equal shares in, identifying, 676, 677, 679, 680
        fourths, 688–692
        halves, 682–686
    as flat surfaces, of shapes, 644–648
    taking apart shapes containing, 637–642
Rectangular prisms
    combining, to make new shapes, 655–657, 659
    definition of, 650
    describing, 649–654
    taking apart shapes containing, 661–666
Repeated Reasoning, 135
Response to Intervention, *Throughout. For example, see:* T-1B, T-115, T-137, T-201, T-241B, T-333, T-443, T-539B, T-587, T-615
Review & Refresh, *In every lesson. For example, see:* 8, 76, 210, 278, 372, 444, 514, 588, 624, 692

Rhombus
    combining shapes to make, 627, 628, 633, 635
    definition of, 620
    describing, 619–624
Rows
    in 120 chart, 293–294
    in hundred chart, 404, 406, 410, 413

S

Scaffolding Instruction, *In every lesson. For example, see:* T-5, T-147, T-231, T-337, T-369, T-411, T-505, T-555, T-645, T-689
Shapes, *See also specific shapes*
    three-dimensional
        combining to make new shapes, 655–660
        curved surfaces of, 644–648
        describing, 649–654
        edges of, 650–654
        flat surfaces of, 644–654
        sorting, 643–648
        taking apart, 661–666
        vertices of, 649–654
    two-dimensional
        closed or open, 614–618
        combining to make new shapes, 625–636
        definition of, 614
        describing, 619–624
        equal shares in, fourths, 687–692
        equal shares in, halves, 681–686
        equal shares in, identifying, 675–680
        number of sides, 614–624
        number of vertices, 614–624
        sorting, 613–618
        taking apart, 637–642
Shortest
    definition of, 504
    ordering objects by, 503–508

Show and Grow, *In every lesson. For example, see:* 4, 66, 128, 188, 244, 294, 356, 404, 460, 504

Show how you know, *Throughout. For example, see:* 24, 295, 208, 235, 278, 301, 349, 588

Sides, of two-dimensional shapes
    definition of, 614
    describing, 619–624
    sorting by number of, 614–618

Smaller unknown, compare problems with, 151–156

Spheres
    combining with other shapes, 655
    definition of, 650
    describing, 649–654
    taking apart shapes containing, 661

Squares
    combining, to make new shapes, 627, 635
    combining shapes to make, 632
    definition of, 620
    describing, 619–624
    equal shares in, identifying, 676, 677, 679, 680
        fourths, 688, 689, 691, 692
        halves, 682, 683, 685, 686
    as flat surfaces, of shapes, 644–648
    taking apart shapes containing, 637–642

Start, unknown
    add to problems with, 127–132
    take from problems with, 139–144

Straight sides, of two-dimensional shapes
    describing, 619–624
    number of, 614–618

Structure, *Throughout. For example, see:* 47, 50, 53, 73, 109, 129, 141, 171, 201, 260, 295, 384, 423, 467, 683

Subtraction
    of 0 or all, 71–76
    of 1, 77–82
    using addition to subtract strategy, 113–118
        within 20, 249–254
        of tens, 439–444
    using bar model, within 20, 281, 282
    using "count back" strategy, 107–112, 434, 437
        within 20, 243–248, 280
    using "count on" strategy, 250–254, 440, 443
    definition of, 28
    using "get to 10" strategy, 261–266, 426
        subtracting 9 in, 255–260
    solving problems using
        within 20, 279–284
        how many fewer, 39–44, 151–156
        how many more, 34–38
        smaller unknown, 151–156
        take apart, connected with put together, 51–56
        take from, 27–32
        take from, with change unknown, 133–138
        take from, with start unknown, 139–144
    of tens, 427–432
        using addition to subtract strategy, 439–444
        using mental math, 409–414
        on number line, 433–438

Subtraction equations, *See also* Subtraction
    completing fact families with, 169–174, 520
    how many fewer, 40–44
    how many more, 34–38
    take apart, 52–56
    take from, 27–32
    true or false, 157–162, 267–272, 372, 482
    writing, 28–32

Success Criteria, *In every lesson. For example, see:* T-3, T-71, T-139, T-267, T-323, T-445, T-521, T-583, T-661, T-687

A21

Sums, 10 (*See also* Addition)
    adding zero (0) and, 66
    given, finding unknown addends for, 21–26

Surfaces, 644–648
    describing shapes by, 649–654
    sorting shapes by, 643–648

Symbols
    equal to (=), 10, 373–378
    greater than (>), 356, 373–378
    less than (<), 356, 373–378
    minus sign (−), 28
    plus sign (+), 10

T

Take apart problems, 52–56
Take from problems, solving, 27–32
    with change unknown, 133–138
    with start unknown, 139–144

Taking apart shapes
    three-dimensional, 661–666
    two-dimensional, 637–642

Tally charts
    completing picture and bar graphs with, 559–564
    definition of, 542
    making, 541–546, 559
    picture graphs compared to, 547
    solving problems with data from, 565–570

Tally mark, 542 (*See also* Tally charts)

Ten frames
    for composing numbers, 305, 307, 310
    for finding missing addend, 100
    for "get to 10" strategy, 261–266, 426
        in subtracting 9, 255–260
    for identifying true or false equations, 268, 271
    for "make a 10" strategy, 163–168
        in adding 9, 217–222
        in adding two numbers, 223–228
    for making true equation, 274

Tens (10)
    adding, 415–420
        in adding two numbers, with ones, 459–470
        in adding two-digit numbers, 477–482
        using mental math, 403–408
        to number, 445–450
        on number line, 421–426, 446, 449, 465–470, 484, 490, 493
    in comparing numbers
        10 more, ten less, 385–390
        11 to 19, 355–360
        within 100, 361–372, 374
    in composing or writing numbers, 311–316
        11 to 19, 305–310
        to 120, 341–346
        counting for, 317–322
        in different ways, 335–340
        quick sketches for, 323–328, 360
        in two-digit number, 329–334
    counting to 120 by, 299–304
    definition of, 306
    in "get to 10" strategy, 261–266, 426
        subtracting 9 in, 255–260
    in "make a 10" strategy
        adding 9 in, 217–222
        adding one-digit number to two-digit number in, 471–476
        adding three numbers in, 211–216
        adding two numbers in, 223–228
        finding addends in, 163–168
    subtracting, 427–432
        using addition to subtract strategy, 439–444
        using mental math, 409–414
        on number line, 433–438

Tens place, 306, 329–334

Think and Grow, *In every lesson. For example, see:* 4, 66, 128, 188, 244, 294, 356, 404, 460, 504

Think and Grow: Modeling Real Life, *In every lesson. For example, see:* 6, 68, 130, 190, 246, 296, 358, 406, 462, 506

Three-dimensional shapes
    combining to make new shapes, 655–660
    curved surfaces of, 644–648
    describing, 649–654
    edges of, 650–654
    flat surfaces of, 644–654
    sorting, 643–648
    taking apart, 661–666
    vertices of, 649–654

Time, telling
    on analog clock, 583–606
    on digital clock, 601–606
    to half hour, 589–594
    to hour, 583–588
    to hour and half hour, 595–600

Trapezoids
    combining, to make new shapes, 626, 628, 630, 633
    combining shapes to make, 627, 630
    definition of, 620
    describing, 619–624
    equal shares in, identifying, 676, 679
    taking apart shapes containing, 638, 639, 641

Triangles
    combining, to make new shapes, 625–630, 632, 633, 635
    definition of, 620
    describing, 619–624
    equal shares in, identifying, 677
    taking apart shapes containing, 637–642

True equations, finding number making, 273–278

True or false equations
    definition of, 158
    identifying, 157–162, 267–272, 372, 482

Two-digit numbers
    adding one-digit number to, 471–476
    adding using place value, 477–482
    comparing
        1 more, 1 less, 385–390
        10 more, 10 less, 385–390
        11 to 19, 355–360
    comparing, within 100, 361–366
        using number line, 379–384
        using place value, 367–372
        using symbols, 373–378
    understanding place value in, 329–334

Two-dimensional shapes
    closed or open, 614–618
    combining to make new shapes, 625–636
    definition of, 614
    describing, 619–624
    equal shares in
        fourths, 687–692
        halves, 681–686
        identifying, 675–680
    number of sides, 614–624
    number of vertices, 614–624
    sorting, 613–618
    taking apart, 637–642

Unequal shares
    definition of, 676
    identifying equal shares *vs.*, 675–680

Unknown(s)
    bigger, compare problems with, 145–150
    smaller, compare problems with, 151–156

Unknown (missing) addends
    add to problems with, 45–50, 127–132
    missing or unknown, making 10, 163–168
    part–part–whole model of, 94, 128–132
    put together problems with, 21–26, 51–56
    ten frame for finding, 100

Unknown change
    add to problems with, 45–50
    take from problems with, 133–138

Unknown start
   add to problems with, 127–132
   take from problems with, 139–144

Vertex (vertices)
   definition of, 614
   of three-dimensional shapes, 649–654
   of two-dimensional shapes
      describing, 619–624
      L-shaped, 615, 616, 619, 622
      sorting by number of, 614–618

Which One Doesn't Belong?, 603, 606
Whole, *See also* Part–part–whole model
   definition of, 16
   equal shares in
      fourths, 687–692
      halves, 681–686
      identifying, 675–680
   put together problems for finding, 15–20
   subtraction equation for finding, 139–144
Word problems, solving
   with addition, 489–494
      within 20, 229–234
      add to, 3–14
      add to, with missing addend, 45–50
      bigger unknown, 145–150
      put together, 15–20
      put together, connected with take apart, 51–56
      put together, missing both addends, 21–26
   with subtraction
      within 20, 279–284
      how many fewer, 39–44
      how many more, 33–38
      smaller unknown, 151–156
      take apart, connected with put together, 51–56
      take from, 27–32
Writing, 549, 552
Writing equations
   addition, 3–8
   subtraction, 28–32
Writing numbers
   11 to 19, 305–310
   to 120, 341–346
   counting by tens and ones for, 317–322
   decade numbers (tens), 311–316
   in different ways, 335–340
   quick sketches for, 323–328, 360
   understanding place value in, 329–334

You Be the Teacher, *Throughout. For example, see:* 79, 115, 231, 275, 301, 411, 479, 505, 597, 677

Zero (0)
   adding, 65–70
   subtracting, 71–76

Reference Sheet

Symbols

+ plus
− minus
= equals
> greater than
< less than

Doubles

1 + 1 = 2	6 + 6 = 12
2 + 2 = 4	7 + 7 = 14
3 + 3 = 6	8 + 8 = 16
4 + 4 = 8	9 + 9 = 18
5 + 5 = 10	10 + 10 = 20

Equal Shares

 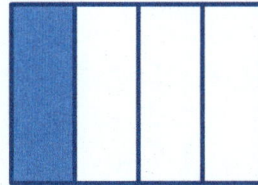 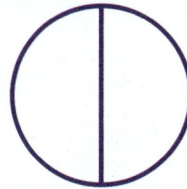 

fourths / quarters   fourth of / quarter of   halves   half of

Time

analog clock   digital clock   An hour is 60 minutes.   A half hour is 30 minutes.

minute hand
hour hand
4 o'clock

half past 4

Two-Dimensional Shapes

triangle
3 straight sides
3 vertices

rectangle
4 straight sides
4 vertices

square
4 straight sides
4 vertices

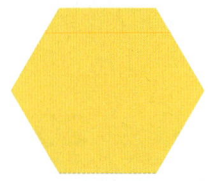
hexagon
6 straight sides
6 vertices

trapezoid
4 straight sides
4 vertices

rhombus
4 straight sides
4 vertices

Three-Dimensional Shapes

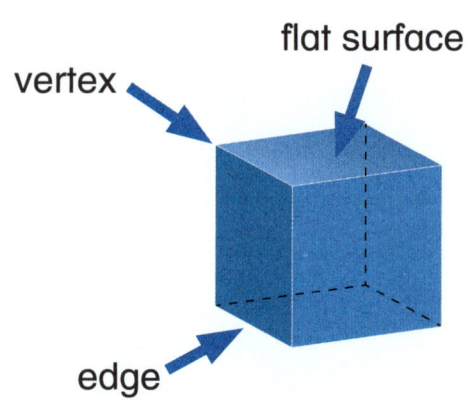
vertex
flat surface
edge

6 flat surfaces
8 vertices
12 edges
cube

6 flat surfaces
8 vertices
12 edges
rectangular prism

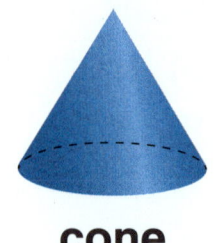
cone
1 flat surface
1 vertex
0 edges

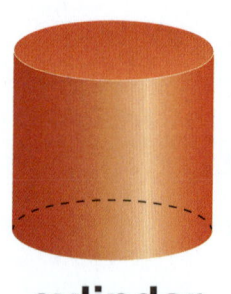
cylinder
2 flat surfaces
0 vertices
0 edges

sphere
0 flat surfaces
0 vertices
0 edges

Credits

Front matter
i Brazhnykov Andriy /Shutterstock.com; **vii** Steve Debenport/E+/Getty Images

Chapter 1
1 Rike_/iStock/Getty Images Plus

Chapter 2
63 Liliboas/E+/Getty Images; **80** Aratehortua/Shutterstock.com

Chapter 3
125 Nastco/iStock/Getty Images Plus, taratata/iStock/Getty Images Plus

Chapter 4
185 macrovector/iStock/Getty Images Plus

Chapter 5
241 DiyanaDimitrova/iStock/Getty Images Plus

Chapter 6
291 shutter_m/iStock/Getty Images Plus

Chapter 7
353 keita/iStock/Getty Images Plus

Chapter 8
401 PaulMichaelHughes/iStock/Getty Images Plus

Chapter 9
457 FatCamera/E+/Getty Images

Chapter 10
501 muchomor/iStock/Getty Images Plus

Chapter 11
539 PeopleImages/iStock/Getty Images Plus

Chapter 12
581 kali9/iStock/Getty Images Plus

Chapter 13
611 Jon (https://commons.wikimedia.org/wiki/File:Ultimate_Sand_Castle.jpg), „Ultimate Sand Castle", https://creativecommons.org/licenses/by/2.0/legalcode

Chapter 14
673 mediaphotos/iStock/Getty Images Plus

Cartoon Illustrations: MoreFrames Animation
Design Elements: oksanika/Shutterstock.com; icolourful/Shutterstock.com; Valdis Torms